Av

WITHDRAWN

EP 8705 24.15

analog days

analog days

THE INVENTION
AND IMPACT OF THE
MOOG SYNTHESIZER

Trevor Pinch and Frank Trocco

HARVARD UNIVERSITY PRESS
Cambridge, Massachusetts, and London, England

2002

For Annika and Benika
and
For Emmett and Zaela

Library of Congress Cataloging-in-Publication Data

Pinch, T. J. (Travor J.)
Analog days : the invention and impact of the Moog synthesizer /
Trevor Pinch and Frank Trocco,
p.   cm.
Includes bibliographical references and index.
Discography: p.
ISBN 0-674-00889-8 (alk. paper)
1. Moog synthesizer.   I. Trocco, Frank.   II. Title.

ML1092 .P56   2002
786.7'419—dc21   2002027257

Foreword

ROBERT MOOG

MUSICAL INSTRUMENT DESIGN is one of the most sophisticated and specialized technologies that we humans have developed. Even the drums and pipes of our distant ancestors were among the most highly developed artifacts of their time. More recently, bowed and reed instruments were assembled from unlikely combinations of materials, each of which was meticulously shaped as a component of a complex structure. Among products that matured in the industrial manufacturing environment of the eighteenth and nineteenth centuries, the piano and the saxophone are unique both in the ingenuity of their design and the precision of their manufacturing processes. Thus, when we speak of musical instruments today, we understand that we are talking about precisely made and finely tuned objects.

On the other hand, musical instrument design has always been at the fringe of technology, far from mainstream practices that stress ease of manufacture, predictability, and economy. Materials such as the woods, glues, varnishes, and catgut of string instruments or the alloys used to make cymbals are selected for properties that defy objective specification; and component pieces, like the neck of a cello or the body of a bassoon, are contoured with organic complexity. Testing and adjustment are largely a matter of human judgment, rather than the application of rulers and gauges. In fact, some of the finest musical instruments are so special and idiosyncratic that nobody has ever learned how to replicate them exactly.

How can it be that musical instruments are both sophisticated technological devices and quirky artifacts that often seem to border on the irrational? I believe that the answer lies in how musical instruments are used. Music-making requires both the musician and the listener to function at the very limits of their perceptive and cognitive capabilities. Therefore, a musical instrument has to be as effective as possible in translating the musician's gestures into the sonic contours that he is envisioning. When he performs, the musician feels his instrument respond as he hears the sounds that it produces. In terms of modern information theory, the musician-instrument system contains a multiplicity of complex feedback loops, so complex, in fact, that contemporary technology has so far not been able to analyze or characterize the nature of the instrument-musician interaction with precision or completeness. Thus, it is not possible to design a musical instrument by beginning with an objective set of performance specifications. Rather, a musical instrument design usually begins with a designer's intuition. In some manner, this intuition suggests to the designer that a certain arrangement of materials will result in an instrument with desirable sound and response characteristics.

Now we get to a tricky question: Where does the intuition come from? Does one attend a major university to learn it, or study reference books? Does one pick it up from a teacher or master? Can you develop it from experience, just by experimenting? How about learning from one or more musician-collaborators? Are you perhaps born with it? The answer to all these questions, I believe, is "Yes, to some extent." They all may contribute, but no one source accounts for all the intuition you need to make good musical instruments. Well then, what else is there? I believe that ideas and concepts permeate our universe and our consciousness, forming what might be called a "cosmic network," and some of us are adept at noticing them and drawing on them. This is not something you learn about in Engineering School. In fact, modern science is just now, through the work of the biologist Rupert Sheldrake and others, addressing the question of how some of us seem to be aware of events at some distance in space and time. This can explain why technical innovations frequently seem to pop up simultaneously in different places. As you read through this book, you will come across several clear examples of the phenomenon of "shared intuition."

Electronic musical instrument technology during the past century has developed through the contributions of many intuition-inspired innovations. At the beginning of the twentieth century, even before the invention of the vacuum tube, the patent attorney and inventor Thaddeus Cahill envisioned a music production and distribution system in which tones were produced by 15-kilowatt electrical generators and distributed over wires similar to telephone lines. With investors' backing, Cahill actually installed such a system in midtown Manhattan. Known as the telharmonium, his system was not a commercial success, but it did foretell the development of the Hammond organ, the electronic music synthesizer, and Muzak. Just a few years after the introduction of the triode vacuum tube, Leon Theremin noticed that whistles from an improperly adjusted radio could be varied by hand motion. From that, he proceeded to develop the space-controlled electronic musical instrument that bears his name. (By the way, Theremin was also the first to develop color television, during the same period that he did his groundbreaking work with electronic musical instruments.) Another early visionary, Maurice Martenot, used circuitry similar to that used by Theremin to design a strikingly innovative keyboard-controlled instrument. Throughout the 1930s and continuing after the Second World War, dozens upon dozens of innovators developed novel electronic musical instruments of all sorts. As electronic technology has itself advanced, the cosmic network has constantly hummed with ideas for new devices that musicians could play.

Few of the early electronic music innovations such as the trautonium, the hellertion, the crea-tone, the oscillion, and the emiriton have become widely accepted. In contrast, today's popular music simply would not exist without the music technology of the past half century or so. Why have most early electronic musical instruments fallen into obscurity, while many recent developments such as the keyboard synthesizer, the phaser, and the fuzz box have become part of the growing electronic musical instrument industry? Rapidly evolving electronic technology is only part of the answer. The complete answer must take into account the evolution of the cultural environment in which we are immersed. Just as a musician interacts with her instrument as her music evolves, technology and our culture are constantly interacting as they themselves evolve. The stories in this book, of how synthesizers came into being, provide fascinating and revealing insights into how technical, commercial, and cultural trends shape one another. In addition, I believe you will find that the stories also shed light on the cosmic network, and how it contributes to human creativity and innovation.

FOREWORD

Contents

CONTENTS

Preface

TREVOR PINCH

THE JOURNEY THAT led to this book began in London in 1970, when I was a physics student at Imperial College, London University. I met a group of people gathered around a space-age box that emitted strange noises. *They* were the Electronic Music Society, and *the box* was a synthesizer (a VCS3 made by EMS—a tiny and much cheaper instrument than its more glamorous American cousin, the Moog). I fell in love with it, and the sixties finally caught up with me.

I played guitar and twelve-bass accordion and moved into a house in Muswell Hill, London, which became a two-year experiment in communal living, technology, and psychedelic sound. I built my own synthesizer from circuits I found in a hobbyist magazine, *Wireless World* (designed by Tim Orr, who later worked for EMS). I loved my homemade synth and played it for years. My first debt is to my fellow Muswell hillbillies Phil, Gill, Viv, Roger, Steve, Mark, Caroline, and Rashmi. Rave on wherever you are.

With the harshness of the seventies and the lure of a new career as a sociologist of science, I put my synth aside. In the early eighties, inspired by David Revill, one of my sociology students who was also an electronic music composer and friend of John Cage (he went on to write a biography of Cage, *The Roaring Silence*), I rediscovered the range of sounds in my old analog synth. It impressed me that my instrument still had something to offer in the digital age, but as it aged I was increasingly reluctant to fire it up—transistors burnt out and my soldering skills were getting rusty. In 1990 when I moved to the United States and to a new job at Cornell University, I crated it up. It now sits unused in my basement.

A moment of epiphany came when I discovered that the Moog synthesizer had been invented in Trumansburg, not far from my new home. The full story had never been told, and a sabbatical at Cambridge University in 1995 gave me time to read into the topic and write my first paper on the history of the synthesizer.

In 1996 Frank Trocco joined me at Cornell from Union Institute to study for his Ph.D. Frank was from my generation. He too had a background playing accordion (button), had traveled round the States interviewing and taping traditional and old-time musicians for the Library of Congress Folk Archives, and had spent much time with the Navajo. He needed an internship and was happy to work with me on the "synthesizer project." At this stage we funded the research from our own pockets or by bootstrapping onto other projects.

We were fortunate to be located at Cornell, where Bob Moog had been a graduate student. Such is upstate life that several of the people from the early days are still living in the area. Bob Moog is, of course, central to our story. But no true innovation comes from one person alone. Many people who are less well known took part in the excitement, successes, and setbacks of the early days. Some had never told their story before, while for others the events they narrated seemed as near and familiar as yesterday.

Tracking down the people who have left the Ithaca area has not been easy. By luck (and thanks to Danny Sternglass) we found Jim Scott—an engineer who contributed to the development of the Minimoog. He was living in a trailer on the Navajo reservation awaiting the fallout from Y2K. Bill Hemsath, the engineer who made the first Minimoog prototype, had also vanished in mysterious circumstances. We were confidently told we would never find him and we had almost given up hope when we did what, with hindsight, should have been obvious—an Internet search. He was the very first hit! Bill now works for an electronics company in Dallas, Texas.

In the summer of 1996 I presented initial research findings at the ICOTECH conference on the history of technology and music in Budapest, Hungary. I am exceedingly grateful to Hans-Joachim Braun for organizing this meeting. During the conference outing—a boat trip on the Danube, with Strauss's "Blue Danube" playing over the ship's loudspeakers—I was approached by someone who said he liked my paper. He was Art Moella, Director of the Lemelson Center for the Study of Invention and Innovation at the National Museum of American History, Smithsonian Institution. This was the beginning of a very fruitful association with the Lemelson Center, which funded our research (now formally carried out in collaboration with the Division of Cultural History, National Museum of American History, Smithsonian Institution).

With this support, we were able to widen the net and interview many more engineers and musicians. As we started to come across old synthesizers, the curators at the Smithsonian asked us to help them build up their synthesizer collection. We have spent a lot of time over the past five years in basements and attics. In spring 2000 we helped organize a special exhibit, panel, and concert at the Smithsonian on the history of the synthesizer. This event brought together a few of the people featured in this book. We thank Art Moella, Joyce Bedi, and Claudine Close at the Lemelson Center and Jim Weaver, Cynthia Hoover, Gary Sturm, and Howard Bass at the Smithsonian Institution for their continuing support and enthusiasm.

This project would not have been possible without all the engineers and musicians who so willingly and generously shared their time and memories through interviews (see Sources). We are especially grateful to David Van Koevering for giving us access to his personal archive and record collection and to Brian Kehew for letting us sift through his unique collection of synth memorabilia, including an invaluable archive of correspondence between Bob Moog and his East Coast sales rep, Walter Sear. We thank Vivian

Perlis and the Yale School of Music and Library for permission to use its Oral History of American Music. Dave Kean of the Audities Foundation, Calgary, let us photograph the best analog synth collection in the world; he and Mark Vail have been a gold mine of information.

The following people have all helped us at many different stages in many different ways: Wiebe Bijker, Karin Bijsterveld, Hans-Joachim Braun, Michael Century, Harry Collins, Peter Dear, Michael Dennis, Park Doing, Mark Elam, Chris Finlayson, Simon Frith, Claudia Fuchs, Karta Iglesias, Natalie Jeremijenko, Ulrik Jorgenson, Peter Karnøe, Beth Kelly, Ron Kline, Roger Luther, Lewis McClellan, Marc Perlman, Richard Rottenburg, Susan Schmidt-Horning, Otto Sibum, Meredith Small, Knut Sørensen, Becky Van Koevering, and Anne Warde. I would also like to thank my colleagues and graduate students in the Department of Science and Technology Studies at Cornell for their continued encouragement and support. A six-month spell at the Society for the Humanities, Cornell University, provided a test lab for some of the ideas in this book. The following institutions all hosted me during different stages of the research and writing of this book: Department of History and Philosophy of Science, Cambridge University; Institute for Technology and Society, Danish Technical University; Faculty of Cultural Studies, University of Maastricht; Max Planck Institute for the History of Science, Berlin; and Faculty of Cultural Studies, Viadrina European University, Frankfurt (Oder). Parts of this book have been presented to many different audiences at numerous talks — I thank them for their participation.

Frank and I did nearly all the interviews together, and Kate Marrone patiently transcribed our tapes, for which we are grateful. The manuscript as a whole was read by Pablo Boczkowski, Brian Kehew, Roger Luther, and Bob Moog. They provided incisive comments and feedback and corrected many mistakes. Parts of the manuscript were read by Don Buchla, Malcolm Cecil, Suzanne Ciani, David Cockerell, Keith Emerson, Bernie Krause, David Van Koevering, and Robin Wood, who all provided additional corrections. Debbie Van Galder helped prepare the final manuscript. All writers depend upon editors, and we were lucky indeed to work with Michael Fisher, Lindsay Waters, and Susan Wallace Boehmer at Harvard University Press. Michael and Lindsay backed this project from the outset and coached us as we got the manuscript into shape; Susan's attention to detail and craft with words improved the final manuscript immeasurably, for which we are grateful.

Special thanks are due to David Borden, Brian Kehew, Mark Vail, David Van Koevering, and Jon Weiss, who became our personal guides and confidants on this journey. Our biggest thanks of all go to Bob Moog, not only for inventing his wonderful machine but also for somehow finding time, in between running his business and his numerous engagements, to be interviewed several times and to help us in his usual good-humored way with what must have seemed liked endless inquiries over matters of detail.

Finally, I wish to thank my wife, Christine Leuenberger, and my two daughters, Annika and Benika, for having put up with so much "weird, hippy music" over the years. Kids, you can go back to surf music now!

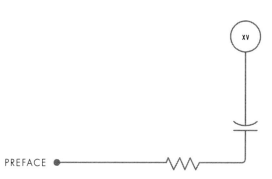

An examination of more recent phenomena shows a strong trend toward spray cheese, stretch denim and the Moog synthesizer.

    FRAN LEBOWITZ *Metropolitan Life* (1978)

Holidays & Salad Days
And Days of Moldy Mayonnaise

    FRANK ZAPPA "Electric Aunt Jemina"
        Uncle Meat (1967)

Introduction: Sculpting Sound

IT WAS LATE SPRING 1996 when we finally tracked down Jon Weiss at Interlaken, a remote hamlet north of Trumansburg, upstate New York. Jon had been a studio musician in the heyday of Moog's Trumansburg factory. Long after all the workers had gone home, he would stay on in the little studio at the back of the factory, composing electronic music deep into the night. He became so adept at manipulating the vast array of knobs and wires that everyone turned to him when a new musician needed a demonstration of the instrument. He was famous in the company as the guy who had taught Mick Jagger how to play the synthesizer. Rock musicians referred to him as The Man from Moog.

Driving through Trumansburg, we passed the storefront on Main Street where it all had taken place. The sign "R. A. Moog Co." had hung proudly outside the building long after it was abandoned by Moog in 1971. Downstairs had become a bar; the new owners had spent years scraping splatters of solder off the wooden floors. A few wrong turns later we finally found Jon's house. We were in the middle of nowhere, just off Route 96, amid rugged, open countryside sweeping down to Lake Cayuga. It was early evening.

Jon was at the door to greet us—a wiry guy with a shock of black hair and a beard. He looked not unlike Jerry Garcia of the Grateful Dead. He now worked for a local garage, specializing in restoring VW bugs. As we sat

Figure 1. Trumansburg, 1971

down in his living room, we could not help but notice the thick, black grease outlining his fingernails.

Jon had something to show us. He produced a black and white photograph of three men playing a Moog synthesizer (see illustration opposite). We recognized it as the Moog Series 900—the standard synthesizer of the day—with its patch wires and multiple knobs. The picture had been taken around 1969 in the Trumansburg studio. In the foreground, with his hands on the keyboard and pens in his pocket protector (as always), was Bob Moog. An earnest-looking young man wearing spectacles was seated in the middle. His right hand was playing a keyboard, his left was reaching above him to adjust a knob as if in salute to the giant synthesizer. That was Jon. He had trained as a violinist, leaving Antioch College to join the Moog company. The third person in the photograph was a customer, Frank Harris, who just happened to be there that day.

At first Jon was hesitant to talk; he could scarcely believe that anyone was interested in what he had been doing thirty years ago. We broke the ice with our own stories about that period. He wanted to know what we knew. He wanted to make sure that we would appreciate what he would tell us. As we settled in for the evening, he introduced us to his son and his wife, Terry, a local hairdresser. A six pack appeared.

We went through our list of interview topics: how he had become involved with the Moog company; who else he knew; how he used the synthesizer, and

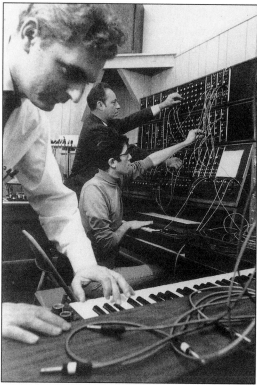

Figure 2. Moog studio, 1969: Bob Moog (left), Jon Weiss (middle), Frank Harris (right)

so on. Our conversation was like an aircraft sweeping low over the sea looking for wreckage. Back and forth we weaved, each pass bringing more details to our attention.

He described some of the excitement and chaos of those early days when synthesizer concerts turned into "happenings." He told us about the way he composed his own electronic music—like a sculptor, he molded sound into new forms—and how he had little use for the keyboard. He told us about his close relationship with Bob Moog, who had been like a father to him, and how Moog had always tried to learn from his musicians, changing the synthesizer to adapt to their needs. He told us about the people who shared his vision of electronic music, such as Sun Ra, the legendary space-jazz musician whose synthesizer never worked as the engineers intended but who made music that was "fabulous." For Jon it was a time of exploration—new sounds, new consciousness, new politics, and new relationships. It was, in a word, the sixties. He told us how his vision of electronic sound as a form of sculpture had not been realized; how the synthesizer had slowly turned into a glorified electric organ on which to play prepackaged sounds. With a wry smile, he announced that today he preferred to listen to acoustic music.

Jon brought up the story of his visit to London in the summer of 1968 to teach Mick Jagger the synthesizer. He ended up living with Jagger for a full month and in that period tasted the rock star lifestyle—riding in Keith Richards's Bentley and partying with the Beatles. He helped Jagger use the synthesizer for the cult movie *Performance* (1970). The Stones came to depend on him for his American know-how about sound technology, loudspeakers, and so on; by the end of the visit they invited him to stay on as their equivalent to Magic Alex, the famed technical guru at Apple whom John Lennon had befriended. But Jon already had a special relationship—with Bob Moog. Synthesizers were his first love, so he went back to Trumansburg. Jon told us how over the years he had thought long and hard about his decision to return. We glanced at each other and at the black grease around his fingernails.

It was getting late. We moved down into the basement. There it stood in the corner—a shadowy presence, row upon row of knobs, patch wires dangling like spaghetti, the ghostly modules of a long-grounded spaceship. Jon had kept his synthesizer, customizing it with new modules, including some from the West Coast inventor Don Buchla. He lovingly described the dif-

Figure 3. Jon Weiss's basement, 1996

ferent bits and pieces and what they all did. Although no longer a working musician, he could never bring himself to part with his synthesizer. He had cared for it all these years—a reminder of what might have been, of what he had lost as well as what he had found.

He hit a switch. Lights flashed. Jon patched in wires and adjusted knobs. The sound of an oscillator grunted into life. He soon had the sequencer set up, and a repetitive pattern of sounds flashed by faster and faster, the tone color changing as the filter came into play, tantalizing like shimmering icicles in the higher frequencies, then cascading downward through the deep resonant tones—the famous fat squelch of the Moog filter. Onward he patched and patched.

It was time to go. At the top of the basement stairs Jon left us with one last image: Woodstock, after the rainstorm in the morning. A lone synthesizer is

5

on stage with two oscillators beating almost in unison, the sound sweeping out over the half million gathered there. Jon: "Those sustained, powerful sounds had never been heard before. It was overwhelming, and it was morning . . ."

◎

Today

It is no longer morning. And now synthesizers are everywhere. They are used in almost every genre of music—from country and western to techno. Japanese multinationals such as Yamaha, Roland, Korg, and Casio dominate the commercial market; the synthesizer has become a truly global instrument.[1] In Sri Lanka, one of the poorest countries on the planet, we have seen a Roland synthesizer played at a beach hotel during a traditional wedding. With electronic dance music dominating the clubs, the driving beat of the synthesizer is once more back in vogue.

In 1964 when Bob Moog and Don Buchla first put together their prototype synthesizers, electronic sounds were limited to a few special effects in Hollywood or to the esoteric music of composers such as Karlheinz Stockhausen and John Cage. Working with synthesizers was seen strictly as a weird and marginal activity. But the revolution in sound that started in Trumansburg thirty-five years ago produced more than just a new musical instrument. Today we are saturated by electronic sounds. Gadgets scream, beep, and growl at us, signaling that our cars have been stolen (or more likely not stolen), that our computers have booted up, or that someone on a TV show is about to become a millionaire. The sound cards in our computers use a technology that is directly descended from the first commercially successful digital synthesizer, the Yamaha DX7 produced in 1983.[2] The patent on the form of synthesis used, known as frequency modulated (FM) synthesis, was for years among Stanford University's highest earning intellectual properties.

The advent of the synthesizer is one of those rarest of moments in our musical culture, when something genuinely new comes into being. Although ingenious inventors have come up with many ways of making and controlling sound and created many precursors to the synthesizer, nearly all of these inventions have remained merely museum oddities.[3] When one thinks of the important new instruments of the twentieth century, one thinks of the electric guitar. The synthesizer is the only innovation that can stand alongside the electric guitar as a great new instrument of the age of electricity. Both led to new forms of music, and both had massive popular appeal. In the long run the synthesizer may turn out to be the more radical innovation, because, rather than applying electricity to a pre-existing instrument, it uses a genuinely new source of sound—electronics. It is the radicalness of the instrument that has allowed the synthesizer to evolve into the digital age. By using a purely electronic source of sound, a synthesizer (now available as just one chip) can be built into any electronic device where sound is needed. The form that today's synthesizers take means they are the instruments par excellence of the digital age. Behind every MP3 file downloaded from the Internet lies some form of synthesizer.

But this book is not about the digital age. Rather, we tell the story of how this all came to be: how the electronic music synthesizer was invented, the people who invented it, and its impact on music and popular culture.[4] We write about what we call the "Analog Days"—the early years of the synthesizer, between 1964 and the mid-1970s, when the technology was analog.[5] Rather than using 1s and 0s, the bits of the digital age, the early sounds were made with continuous variables such as changing voltages.

Robert Moog is the best known of the synthesizer pioneers, and much of this book is about him and the Moog synthesizer. But Moog was not the only inventor to develop a synthesizer in the early 1960s. Working out of a West Coast storefront around the same time, with a similar technology but a totally different vision of electronic music, was Don Buchla, an experimental musician and instrument designer. Buchla, unlike Moog, rejected the use of the conventional keyboard to control this new source of electronic sound. In the end, keyboards won out, at least for most uses.[6] The synthesizer, by the mid-1970s, had become a portable instrument with a keyboard controller. Why Moog's vision triumphed is one of the questions this book sets out to answer.

We would probably not have heard of the Moog synthesizer at all if it had not been for Wendy Carlos, who laboriously assembled electronic music in the studio and produced the sensational album *Switched-On Bach* (1968). This record made Moog and Carlos famous, was responsible for introducing many other musicians to the Moog, and led to a whole genre of "switched-on" records, including *Switched-On Bacharach* (1969), *Switched-On Nashville* (1970), and *Switched-On Santa* (1970). But rock and pop music was where the Moog synthesizer found its true home. Groups like the Byrds, the Doors, and the Beatles used their Moogs as part of the sixties' search for new psychedelic sounds. We also pay attention to lesser known people, such as the few women synthesists who worked in this predominantly male world.

During the early years of the synthesizer, a pivotal part was played by the Minimoog, produced in 1970. One of the first portable keyboard synthesizers, the Minimoog has since become a classic. In the United States it was the first synthesizer to be sold in retail music stores and to be bought in significant numbers by young rock musicians. When Bob Moog was awarded the 2001 Polar Prize by the King of Sweden for his contributions to music, it was for his invention of the Minimoog.

The Moog would have remained a studio instrument, an oddity, if it were not for the efforts of musicians like Keith Emerson, who used it for live performances with his progressive rock band Emerson, Lake and Palmer. Eventually the synthesizer reached mainstream black music, most notably when Stevie Wonder took it up in the early 1970s and introduced the Moog sound to yet a new audience.

The Moog was put to innovative uses in making radio and television commercials and sound effects and electronic scores for films. Other companies, such as ARP (pronounced "arp") in the United States and EMS in the UK, started synthesizer production. By the mid-1970s ARP had become the dominant manufacturer, with a 40 percent share of the $25 million market.[7] ARP synthesizers were featured in the blockbuster sci-fi movies *Star Wars* and *Close Encounters of the Third Kind* (both 1977). With their spectacular range of sound effects, EMS synthesizers were favorites among European art and progressive rock bands. They were used famously by Brian Eno and Pink Floyd.

As we follow the evolution of the synthesizer from the acid dawn of the sixties through the summer of love and into the harsher commercial world of the seventies, we will see that not only did the synthesizer change but so too did the range and sorts of sounds it made. Today in the digital world there is a longing to get back to what was lost; an "analog revival" is taking place. Synthesizers that were invented thirty years ago are still manufactured unchanged and are purchased by modern musicians for many genres of music, including electronic dance music, where analog sounds are much sought after. Old or "vintage" synthesizers command high prices, and Bob Moog has become a cult hero for many young musicians.[8] We end the book by asking why analog days are here again.

◯

Technology and Culture

In the chapters that follow, we will show that technology and cultural practices are deeply intertwined. Often academics separate the two (the two-culture problem), with the result that culture is examined in one corner while technology is analyzed in another. Just as the development of the synthesizer demanded collaboration across cultures, among engineers, musicians, and salespeople, in our story, too, we want to reintegrate machine and music, technology and culture. In the practice of the peoples' lives who created this new industry, they were constantly interwoven.

The analysis of sound, music, and musical instruments in *Analog Days* has been deeply informed by our own background in the new interdisciplinary field of science and technology studies.[9] We conceive of science and technology as sets of practices, discourses, and material artifacts that have evolved over human history and that can take on new forms in different social, cultural, and historical contexts.[10] Although having global import and use, science and technology are always produced, maintained, and consumed locally. There is no royal road to scientific method and no certain impact of technology, although many would claim otherwise.

Similarly for us, the production and consumption of sound, music, noise, and silence involves sets of practices, discourses, and material artifacts that vary from context to context. Some kinds of music and musical practices—for instance, the so-called classical repertoire—show remarkable stability. Others—for instance, electronic dance music and its constantly mutating brands (techno, house, jungle, trance, garage, and so on)—show almost continuous change.[11] For us this is not evidence of some musical essence or lack of it but a social phenomenon itself in need of explanation. We take it as axiomatic that no one canon of musical appreciation is to be elevated over all others. What counts as music and what counts as noise is contested territory.

Wittgenstein famously argued that the way to understand language is from its use.[12] Similarly, the way to understand musical instruments is not from their essences—what their theoretical possibilities are—but from the way people who actually make the music put them into practice. Although instrument designers may have dreams and aspirations for the sorts of music to which their instruments can be adapted, the way to find the meaning of an instrument is in its use by real musicians—in state-of-the art recording studios and home basements, on the stage and on the road.[13]

And let us not forget the importance of the synthesizer in American popular culture. Bob Moog and Don Buchla are not as well-known as Bill Gates of Microsoft or Steve Jobs of Apple Computers. But working at a similar time from small storefronts and garages, they too produced an electronic revolution—in the way music is produced and consumed. The development of the synthesizer will, we think, eventually come to be seen as one of the most significant musical moments of the twentieth century.

The paradox of history is that significant events are often recognized long after they occur, when it may be too late to recapture what went on and why. In hindsight, all arrows seem to point one way, and we forget that the picture was much more confusing at the time, with myriad road signs pointing in many different directions. We try to avoid hindsight. By tracking down and interviewing the early pioneers—engineers, musicians, and other users—we have tried to recreate the enthusiasm and uncertainties of *what it was like back then*, before anyone knew what it would be like now. We use the pioneers' own words to describe their visions, their excitement, and their disappointments.

We see our own task in writing this history as being akin to the practice of analog synthesis. Our sources of sound are the stories we recorded and discovered in texts. We have filtered the stories to bring out certain themes and have muted others. We have shaped our account, giving it narrative structure, in the way that synthesists shaped sound. We have, on occasions, fed the stories back to the participants and hence produced yet new versions of events. Sometimes when stories do not match up, rather than get rid of the inconsistencies, we have allowed the discordances to remain. If we had chosen another configuration of quotes, we are quite sure we could have produced a rather different history. Analog synthesists tell of producing beautiful pieces of music that vanished when they tried to reassemble the patches the next day—the early synthesizers seldom sounded the same from day-to-day, from patch to patch. Our story has inevitably also involved loss, and we are acutely aware of this. There are silences, and noise is everywhere.[14] We have fine-tuned and patched as best we can.

If we have told the story well, we will have brought to life the part played by one machine in shaping our culture, and the part played by our culture in shaping this one machine. The paradox is that our story is about sound—and words alone can never express what it was like to hear for the first time the beat of a pair of oscillators through a big sound system, in a vast arena, in the early morning, after the rain. Sound is the biggest silence in our book.[15]

1

Subterranean Homesick Blues

Johnny's in the basement
Mixing up the medicine.
Bob Dylan

You just couldn't keep Bob Moog out of the basement. It was his space. Or, to be more accurate, it was his dad's space. The house at 51–09 Parsons Boulevard, was a typical Cape Cod–style bungalow in a middle-class residential section of Queens, New York. In postwar America, when keeping up appearances was everything—especially upstairs—Bob Moog was not the sort of kid to keep up appearances. He was shy, awkward, with a mop of wavy hair that his mother insisted should be combed. He wanted out; and for him, out was down—downstairs, in the basement, where his dad, an engineer for Con Edison, had set up a dream workshop.

Here's how Bob saw things: "He [my father] knew a little bit about electronics but not enough to actually design something. But he got me going, taught me how to use a soldering iron; and I could use his shop and I loved working with him, it was solace there. You know when I was in my mother's presence, then I had to worry about was I practicing the piano enough."

Bob's dad was one of the first amateur radio operators in America.[1] Along with a pile of radio equipment, the basement contained a full range of stationary power tools and a huge selection of hand tools. Every evening after his dad came home from work, and after Bob had dutifully practiced piano and done his homework, he would join him in the basement. Bob did what any smart kid who wasn't interested in girls or street fighting did: he escaped to a world that kept girls and the street out. "I think it was primarily my nature, the patch cords in my brain. I was just a goofy, shy, unsocial kid. I always seemed to be out of it, no matter what social circle I landed in . . . and my public school was lower-class Catholic . . . These kids were forever fighting and beating each other up . . . and I couldn't relate to it."

Bob may not have been any good at fighting, but he was good at science. He won a place at New York's Bronx Science, one of the best high schools for science anywhere in the United States. But there he still felt socially isolated, "and here are all these super-vain, loquacious, garrulous Jews and I was out of that too because I was a shy kid and all these kids had fathers who were lawyers, they were businessmen, they talked smoothly and urbanely and I never saw my father talk that way."

Still, Bob (who was born in 1934) had some things going for him. To start with, he was not alone. In postwar America the electronic hobbyist craze was in full swing. Whole swathes of magazines, with titles like *Radio Craft* and *Radio News*, brought hobbyists a new project with each issue: a crystal radio, a one-note organ, a garage door opener, and so on. There were cheap war-surplus and industrial-surplus parts aplenty, and on the way home from school Bob would often stop by "Radio Row" (situated around Fulton, Dey, and Cortlandt streets) to pick up vacuum tubes and boxes of capacitors. His father would bring home scrap metal from work for making panels, chasses, and so on.

And it wasn't just in America. In Britain it was the same. One of us (Pinch) bought his first short-wave receiver from a war-surplus store in a provincial city in the UK. It was an R1155 set that had been stripped from a Lancaster bomber and had the words "Eager Beaver" etched above the giant tuning dial on the front panel. This receiver could tune in all sorts of illicit stuff, like Radio Havana. QSL cards (sent out by short-wave radio stations to listeners) could send your parents into a tail spin—"that's communist propaganda, son." As a bonus, you could also imagine you were bombing Berlin.

If your interest was in making electronic sound effects, the surplus stuff was invaluable. Synthesizer pioneer Don Buchla told us how the San Francisco Tape Center, one of the main venues on the West Coast for making electronic music in the early 1960s, used war-surplus gun sights and test equipment. Herb Deutsch, an experimental composer who had a formative influence on Bob Moog, got started with an off-the-shelf square-wave oscillator. Don Preston, who played synthesizer with Frank Zappa and the Mothers of Invention, told us, "I bought oscillators and put them all together, you know [from kits] . . . Even Stockhausen had to make a lot of stuff that he did in the early days."

◎

Bob's First Love: The Theremin

One hobbyist project that captured Bob Moog's imagination was building an electronic musical instrument called the theremin.[2] Named after its Russian inventor, Lev Termen (Leon Theremin), this is the weirdest instrument in the history of electronic music, possibly the weirdest instrument of any sort, ever. The sound is similar to that of a ghostly, wailing human voice. And like the way we control our voices, by continuous movements of the larynx, tongue, and mouth, the sound of the theremin is controlled by continuous human movements—that of the hands or fingers moving through air. The sound always seems to be in motion. Unlike any other instrument, the theremin does not have a tangible solid controller against which you can bash your fingers—no physical resistance or feedback at all. You control the sound by waving your hands near two antennas, one for pitch and one for loudness. Without a moving hand, there is no sound.

The sound is visceral, and people seem to have a primeval connection with the theremin. On seeing one for the first time, they often react by wav-

14

ing their own hands near it to produce sound. It's like first learning to use your own voice: you believe that if you only worked at it a little bit harder you could get it to work *for you*. But it's damned hard work. The theremin is a notoriously difficult instrument to play because of the lack of any physical feedback.

Bob Moog's own connection with the theremin goes deep. He made theremins as a boy, and he still makes them today. He probably loves this instrument more than his own invention, the synthesizer. We've heard Bob joke about this, saying that his first love in life was the theremin and on the way to rediscovering his first love he invented the synthesizer: "I made my first theremin when I was fifteen in 1949. It was a hobbyist theremin. It didn't work especially well. And I just fooled and futzed with it."

Like all electronic instruments, the theremin's source of sound is electrical—two high-frequency oscillators that

Figure 4. Bob Moog, age 17, playing theremin at Bronx High School of Science, New York City

beat against each other to produce a lower-frequency audible sound. It is an electrical property of the hand, its capacitance, that actually controls the sound. First known as the "etherphone," the instrument was invented in Russia in the 1920s. Leon Theremin even got to demonstrate it to Lenin. After a spectacular tour of Europe, he came to New York City in 1927 to promote his instrument. It was an immediate sensation, and soon Theremin, with his Russian aristocratic roots, was the doyen of high-society hostesses. He trained pupils like Clara Rockmore to give concerts in Carnegie Hall. In 1929 RCA acquired the rights to manufacture his invention but only made a few hundred. Although marketed as a popular instrument, it was too difficult for most people to play; and with the Depression setting in, RCA lost interest.

Theremin had an ulterior purpose in visiting the United States: he was a spy. When World War II broke out, he was summoned back to Russia, leaving behind mounting debts and his beautiful African-American wife, the dancer Lavinia Williams. Back in the USSR, Theremin was a victim of a Stalinist purge and ended up in a labor camp. He was rescued when the Soviets rediscovered his electronic genius and eventually awarded him the Stalin Medal for inventing the first passive electronic bugging device, which sat undetected for years in a huge American eagle plaque hung over the American ambassador's desk in Moscow—a gift from the Russian people.

Starting in the mid-1940s, the theremin, with only a very limited classical repertoire, took on a new life in Hollywood as a source of weird and scary sound effects. Hitchcock used it, for instance, in *Spellbound* (1945), and it made an appearance in early science fiction movies like *The Day the Earth Stood Still* (1951). It also became the hobbyist's project par excellence. The trickiest part in building a theremin was getting the coils right. These large inductance coils produced high-frequency electric fields. Winding these coils was Bob's specialty. Working with his father, he figured out how to do it better and better.

Bob's obsession with the theremin took various forms. As well as building them, he published an article on the subject in one of the leading hobbyist magazines.[3] Other hobbyists started to contact him. He was on a roll. At age nineteen, he and his father started a small business, R. A. Moog Co., which they advertised to fellow hobbyists and operated out of their basement. At first they sold theremin coils and, later, complete instruments. Although theirs was only a basement business, Bob and his dad were not short on chutzpa. Their leaflets proclaimed that, "because of advanced design, quality control and thoroughness, the musician can own the R. A. Moog theremin with pride and play it with confidence."[4]

Although Bob's parents had musical aspirations for their only son, his early home life was largely bereft of music: "No concerts. My parents didn't have a phonograph, there was no music in the home at all." Bob's grandfather on his mother's side came from Poland, and Bob thinks his mother entertained hopes that he would turn out to be another Paderewski (the famous Polish pianist and politician). He started piano lessons at age four, practiced for several years, and at age eight attended the Manhattan School of Music to take courses in sight singing and ear training. But once at Bronx Science, he began to lose interest. At college, Bob was known to play piano occasionally in a dance band but maintains that this was "mainly to enhance my social clout. By getting books on how to play this style or that style, and by listening, I desperately tried to play piano well, but I couldn't even come close."[5] Despite Bob's own candid assessment, several people have told us that he had talent. He loves listening to music. He once told us, "When there's music on I have to stop. I can't talk, I can't eat, you know, I have to listen."

Meanwhile, Bob was keeping his mother happy in other ways. He was excelling at science. He left Bronx Science to start a joint degree in physics at Queens College and electrical engineering at Columbia University in 1957. At Columbia Bob learned electronic-circuit design and the latest advances in electronics. He had no inkling of his future. The Columbia-Princeton Electronic Music Center, formally established in 1959 housed the room-sized RCA Mark II electronic music synthesizer, built in 1957, but Bob did not once walk across campus to visit it.[6] This was despite having a lab instructor who was technical advisor to Vladimir Ussachevsky, an electronic composer and joint director of the center. For Bob, the world of

17

serious electronic music could have been a million miles away: "I heard vague mention of this weird musician Ussachevsky who was doing something in the basement somewhere on campus."

At college, away from his family, Bob became a little bit less goofy. He found a group of friends through his fraternity, Alpha Phi Omega, and met his first wife, Shirley May Leigh (known as Shirleigh), at a fraternity party. The only thing that stopped Shirleigh from marrying Bob immediately was her wish to complete her own degree in education at Queens College. Within a couple of days of getting it, she followed Bob upstate to Ithaca, where he had a place in graduate school at Cornell University studying engineering physics. They were married in June 1958, and Shirleigh got a job as a teacher.

At Cornell Bob led a double life. By day he was on campus taking courses and working in the labs on his dissertation project in solid state physics. In the evenings and weekends he was running a small business from his landlord's basement in Bethel Grove, outside of Ithaca, making theremins. And he was getting better and better at building them. In 1961, he published an article in *Electronics World* describing a fully portable transistorized theremin.[7] His Melodia theremin was sold either fully assembled or as a $50 kit. He sold over a thousand kits, and with Shirleigh pregnant (their first child, Laura, was born on May 6, 1961), he used the money to supplement his graduate student research assistantship: "To design the kit probably took six months . . . It took us forty five minutes per kit to wrap everything up in boxes and packages . . . I couldn't get it out of my head that maybe I could be a kit manufacturer. If that was all there was to making a lot of money, well fine."

◯

R. A. Moog Co. Moves to Trumansburg

The double life was to continue a while longer. Bob's calling was not the high church of university physics but rather his first love, the basement workshop. In 1963 he rented a storefront, hired a couple of people, and set up R. A. Moog Co. in Trumansburg, eleven miles north of Ithaca. He was going into business as a kit manufacturer.

People who live in Ithaca joke that it is centrally isolated. Trumansburg is just isolated. But isolation has its advantages—it is cheap to live and work there. It is also spectacularly beautiful. The nearby Taughannock Falls is the highest on the East Coast of America. Bob loved the rustic, rural ideal of Trumansburg and eventually bought an old farm house near the falls where he and Shirleigh raised three (Laura, Renee, and Michelle) of their four children; the fourth one, Matthew, was born in the driveway of the farm house in February 1970. The kids loved to skinny dip in the stream that ran past his home. Bob could sometimes be spotted walking on a nearby hill, deep in thought.

Trumansburg was the last place on the planet that you would expect one of the hippest businesses of the sixties to start. Near Appalachia, this small, conservative farming community is close enough to Ithaca to serve as a bedroom community for the occasional Cornell professor and a number of plant workers. Here is how one of the employee's from Bob's factory, Leah Carpenter, described it: "I was raised in Ithaca, on South Hill. My father always said that Trumansburg is a place with five churches and five bars. It is! It's a village, and basically, everybody knows everybody out there."

◯

Enter the Tuba

"How did Moog get from the theremin to the synthesizer?" is a question often asked. The answer is: by way of the tuba, or rather by way of one resourceful player, designer, and manufacturer of tubas, Walter Sear. The tuba with its flatulent sound and large girth has always been an instrument of comic proportions. "Professor" Jimmy Edwards, an English music hall comedian and fifties radio star, played the tuba for laughs. But this joker of an instrument may have had the last laugh. It helped bring into being the one instrument with the potential to render all the others obsolete—the synthesizer.

Trained with degrees in music, sociology, and chemical engineering, Walter Sear has turned his hand to just about everything at one time or another. He has been a musician, salesman, inventor, studio engineer, manufacturer, B-Movie maker, and porno-movie maker. Like Moog, he started out as a hobbyist, building crystal radio sets as a boy. Raised in the Depression, he acquired practical skills in order to survive. Today, he runs an all-tube studio (which he built himself) in New York. We interviewed him in his battered office, surrounded by his movie posters, mementos, and old equipment. Young rockers passed by, obviously in awe. With a cigarette hanging from his lip, a checked shirt open at the neck, and a sardonic sense of humor, he appeared to be a throwback to another age. But Walter Sear is still one cool dude. New York City rockers like Sean Lennon and bands like Sonic Youth record at his studio in search of his fabled analog sound.

Before he got into tubes, Sear was into tubas. He lived and breathed tubas. He blew them in the Philadelphia Orchestra, in an Air Force band during World War II, and later, on returning to his hometown, New York, in the pit at Radio City Music Hall on Broadway. "Four shows a day, six days a week . . . five dollars a show." Sear's Holton tuba finally wore out. Rather than buy a new one, he decided to make his own. He rented a loft, and between Broadway shows he designed his first tuba. He liked what he made and set up a manufacturing plant for tubas using parts made in Czechoslovakia and Belgium.

He first met Bob Moog when Bob was a student at Columbia. He bought a theremin from him that took him two years to learn to play, "between shows at the Music Hall, practicing." Sear had many contacts in the New York music business and arranged to sell Bob's theremin kits. He was in effect Moog's agent: "I dragged them around to different jobs and I'd show the arrangers the instrument."

Walter Sear's tuba business was doing nicely. The major instrument manufacturers "couldn't really be bothered with tubas." He sometimes attended trade shows and educators' conventions, where he would take his tubas along to demonstrate and perhaps generate a few sales. "So, anyway, there was a trade show of the New York State School Music Association [NYSSMA]. I figured I'd go up there [to Rochester] and, you know, show the tuba. And since Bob was in Ithaca, Trumansburg, I said, 'Why don't you come up?' we'll have a little fun and we can show some theremins and stuff." Bob was not going to say no, especially if he could mix fun and theremin sales. He packed up some kits into the trunk of his VW bug and headed for the educators' convention at the Eastman School of Music. It was winter 1963.

At the convention Sear found business anything but brisk: "[Bob] just sort of hung around; worked the aisles, as we say . . . With literature, handing it out . . . Once in awhile we'd put [a theremin] out. That always would attract a crowd and then I'd try and sell the tubas." With its wailing sound, the theremin was the perfect instrument to draw attention. One person to stop by Sear's table was a young experimental composer, Herb Deutsch, who had an interest in electronic music and owned some rudimentary equipment, including a Moog theremin, which he set up in the living room of his Long Island home. He was thrilled to meet Moog himself, as he had no conception of how small Moog's business was: "There was this man standing there with Moog theremins, and, naive as I was, I assumed that Moog theremins were being sold all over the place and that this was a salesman . . . It was Bob and, of course, we started to talk." Moog, too, remembers that happy moment: "Herb Deutsch came around, introduced himself, said, 'I used one of your theremin kits in class to teach sight sing-

21

ing.' And then he said, 'Do you know anything about electronic music?' And I said something to the effect of, 'Sure.' And I really didn't!" They went on talking for hours, and Deutsch invited Bob to attend a concert he was giving in New York City. It was from this conversation that the germ of an idea for a new instrument emerged.

○

A Loft with Electronic Music Sculptures

Moog was at heart an engineer. The main electronic music traditions from the 1950s such as *elektronische musik* in Cologne, associated with composers Herbert Eimert and Karlheinz Stockhausen, and *musique concrète* in Paris, associated with Pierre Schaeffer, were not part of his world. Even America's own burgeoning electronic musicians such as John Cage and David Tudor were almost unknown to him. And certainly he had never experienced anything like the New York experimental music scene of which Deutsch was a part. It was a time in Greenwich Village when artists and musicians mingled, experimenting in new media.[8]

The concert to which Deutsch had invited Moog took place in the Manhattan loft of sculptor Jason Seley. The highlight was a piece of Deutsch's, "Contours and Improvisations for Sculpture and Tape Recorder." Deutsch ran a tape while a percussionist hit Seley's automobile bumper sculptures, which stood six to eight feet high and made an impressive sight. "Besides being visually dramatic, they were incredible percussion instruments, because you could hit them in one place and get one tone, and hit them in another place and get another tone." Moog, who recorded the concert on his portable Berlant tape recorder, had the time of his life: "It was absolutely the most exciting musical performance I had ever seen up until then. I hadn't seen that much, before rock." Deutsch also remembers the importance of that performance: "And it was immediately after that concert that we talked about a synthesizer. We didn't call it a synthesizer . . . what we were talking about is a sort of portable electronic music studio." The need for more compact equipment was something that other composers recognized at the time.[9]

Deutsch was struggling to make electronic music with his square-wave oscillator, theremin, and tape recorder. It was an elaborate and time-consuming process, recording individual sounds and then cutting and splicing tape. Moog was ready to help out. He asked, "You know, what do you want to be able to do, Herb?" Deutsch replied, "Well I want, I want to make these sounds that go *wooo-wooo-ah-woo-woo*." Moog: "He was interested in sounds with moving pitches. He was frustrated by the fixed pitch generators that he was using and the lack of control he had over other aspects of sound."

If there was one thing Moog knew from tinkering with electronic circuits most of his life, it was how to make certain sorts of sounds and how to control them. Ever since he was a kid back in his dad's basement he had worked with sound and knew the shapes of different waveforms and the sounds they made. With the aid of an oscilloscope he used his eyes to see the shape of the waveform; with the aid of a loudspeaker he used his ears to hear the sound of the waveform; and with the aid of a voltmeter he used his hands to tinker with the circuit producing the waveform. The musicians might play the instruments, but the engineers played the circuits, and to play them well they needed to have good eyes, ears, and hands. He knew that sort of stuff "stone dead" since childhood. He went back to Trumansburg and started to tinker.

◎

Voltage Control

It was an opportune time to do "*wooo-wooo-ah-woo-woo*" in new ways. Things were also happening in electronics. Cheap silicon transistors had become widely available, replacing the bulky and expensive vacuum tubes. One newly introduced form of the silicon transistor was of particular interest to Bob. It had an exponential relationship between its input voltage

23

and output current over the frequency range of musical interest (several octaves). Exponentially varying properties are common in music; for instance, the frequency change between two tones is exponentially related to the pitch change (a one-octave increase is a doubling of frequency). Another example is sound intensity, which is exponentially related to loudness.

Moog now had a key insight. Rather than varying the pitch of an oscillator manually by turning a knob or, in the case of the theremin, by moving a hand, he could make the pitch change electrically by using a "control voltage" to vary it. A larger voltage fed into the oscillator as a "control" would produce a higher pitch. This meant that an oscillator could be swept through its total pitch range (several octaves) simply by increasing the voltage. Similarly, a voltage-controlled amplifier could be swept through the complete dynamic range of human hearing. By building "exponential converter" circuits into his devices—circuits that converted a linearly varying parameter like a voltage into an exponentially varying parameter like frequency or intensity—Moog could make these control voltages musically useful. It enabled him to design all his modules around a single standard—the volt-per-octave standard—such that a change of a control input of one volt produced a change in the output pitch of one octave.[10]

Moog and Deutsch were excited by their meetings. They knew they were on to something, even if they didn't quite know what it was. In early February, Moog sent Deutsch the tapes of the concert. The accompanying letter shows Moog promising to have "a good assemblage of 'studio equipment' in a few months."

Around this time, Moog designed a new logo for his company, replacing the one based on a tube, which he had used for the last ten years, with a single musical note surrounded by a circle. He managed to secure a grant of $16,000 from the New York State Small Business Association. Trumansburg, like most of upstate New York, was economically depressed, and so the state was willing to give money to new business ventures. To further his collaboration with Deutsch, Moog invited him to bring his family to Trumansburg for a vacation. And the omens were good — Deutsch too managed to get a small grant for the trip from the Music Department at Hofstra University, where he taught.

○

"Weird Shit!"

Deutsch rented a cabin overlooking Cayuga Lake and drove the two miles up the hill to Trumansburg to see what Bob was up to. What Bob was up to didn't look very impressive — a few transistors wired together, along with a couple of potentiometers. Moog: "I had this little breadboard with three different circuits on it: two voltage-control oscillators and a voltage-control amplifier. They weren't accurate and they weren't a lot of things, but they had the advantage of voltage control. You could change the pitch of one oscillator with the other oscillator. You could change the loudness." Moog compared that breadboard to a hot rod: "It's an electronic circuit that's all hanging out so you can get in and change things quickly. So it's like a hot rod without any body on — everything is sticking out."

Having two voltage-controlled oscillators as opposed to one doesn't sound like very much, but it was a breakthrough. The two oscillators were designed so that the output from one (itself a varying voltage) could be used to control the pitch of the other or the loudness of the signals via the voltage-controlled amplifier. By adding a slowly varying sine wave as an input to an oscillator, a vibrato effect could be obtained. Feeding the same input into a voltage-controlled amplifier could produce a tremolo effect. But this was only the start. Many, many more interesting sonic effects could be obtained by experimenting and feeding back signals that in turn could be used as new controls. This was the secret to making pitches move. The hot rod now was ready to roar.

Moog describes what happened next: "Herb, when he saw these things, sorta went through the roof. I mean he took this and he went down in the

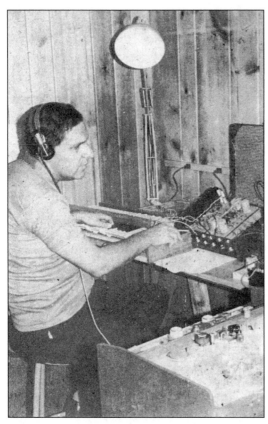

Figure 5. Herb Deutsch with prototype Moog synthesizer, Trumansburg, 1964

basement where we had a little table set up and he started putting music together. Then it was my turn for my head to blow. I still remember, the door was open, we didn't have air conditioning or anything like that, it was late spring and people would walk by, you know, if they would hear something, they would stand there, they'd listen and they'd shake their heads. You know they'd listen again—what is this weird shit coming out of the basement?"

The "weird shit" was historic. It was the first sounds from the very first Moog synthesizer.

○

"Buy Me a Doorbell Button"

Deutsch visited Moog again in July, and they worked together refining and rebuilding the circuits. Since childhood Deutsch had been fascinated by mechanical objects. Though he had no formal technical training, he knew enough about electronics to assemble a theremin kit and dabble with oscillators. Moog, for his part, had enough knowledge of music to understand what Deutsch was doing and how he could be helped. It was the first of many such successful collaborations between Bob and a musician.

They both recall with fondness that early honeymoon period working together. One thing that came out of it was Moog's first version of an envelope generator, which later became a standard device on all synthesizers.

An envelope generator allows the loudness of sound to be structured or contoured to produce, say, the effect of a string being plucked, where the loudness builds up rapidly and then decays away slowly. By this stage, Moog had hooked up an old organ keyboard. Deutsch: "I said, 'It would be great if we could articulate the instrument,' and Bob said, 'What do you mean?' I said . . . 'You play a key, it was on and you lift your finger off and it was off.' . . . And he thought about it and he said, 'Okay, that's easy enough to do.' So he said, 'Listen, do me a favor. Go across the street to the hardware store and buy me a doorbell button.' So I went across the street and I bought a doorbell button for 35 cents . . . and he took out a yellow piece of paper and he started throwing a few formulas and things down."

Moog had found a way, using a doorbell and a capacitor, to store and slowly release a voltage produced at the same time as hitting a key. He soon refined this early design so as to avoid the need to push a separate button with every key press. He put two switches on every key, one to produce the control voltage and the other to trigger the envelope generator. Moog, at the time, had no idea of the significance of what he was doing, "I've always had trouble with differentiating business from hobby . . . It was fun, it was interesting, maybe it would lead somewhere, who knows?"

◎

Endorsement from Myron Schaeffer

Later in that historic summer of 1964 Moog and Deutsch drove up to the University of Toronto Electronic Music Studio to demonstrate what they had made. The composers at the Toronto studio, headed by Myron Schaeffer, were impressed. One of them, Gustav Ciamaga, gave Moog an important idea. He suggested that he should develop a filter module. A filter is a way to remove certain frequencies of sound from a waveform. By attenuating some overtones it is one of the most powerful ways of making sonically distinct sounds. A voltage-controlled filter can be swept to attenuate a changing range of frequencies, making the sound even more interesting. Moog's first filter was little different from a wah-wah pedal on a guitar. It was only later that he designed his famous ladder filter, which became the most distinctive feature of the Moog sound.

There were two crucial ideas embedded in what Moog was doing: first, that voltage control could be applied to an electrical musical instrument, and, second, that the instrument could consist of discrete modules (oscillators, amplifiers, envelope generators, and, later on, filters) that could be wired together in a variety of ways and controlled by the output voltages of the devices themselves. The use of discrete modules mimicked the way electronic composers like Deutsch worked and showed the power of thinking of the synthesizer as a "portable electronic studio."

Another important aspect of Moog's approach was his ability to generate different shaped waveforms from his oscillators. Different waveforms contain different combinations of tones and overtones (all musical sounds consist of a fundamental frequency or pitch and overtones at higher frequencies or pitches). His oscillators produced a sawtooth waveform that is very rich in overtones and makes a bright, full, brassy sound; a triangle waveform that sounds much thinner and purer, like a flute; a pulse wave that produces a nasal, reedy sound; and a sine wave that sounds like whistling. By modulating and then filtering these waveforms, Moog was able to make even more interesting sounds. Because this approach starts with complex waveforms from which overtones are removed, it is known as subtractive synthesis.[11]

There was nothing particularly original in the notion of voltage control or in the design of the circuits Moog employed.[12] Moog's special skill was in drawing the different elements together, realizing that the problem of exponential conversion could be solved using transistor circuitry and building such circuits and making them work in a way that was of interest to musicians. And it was here that his collaboration with Deutsch and other musicians was so important.

Toronto was Moog's first ever visit to an electronic music studio. He was particularly curious because much of the equipment had been designed by Hugh Le Caine, the legendary Canadian physicist who designed electronic music equipment.[13] Moog told us that Le Caine was "like me"—in other words, Bob was not alone.

○

First Sales

With Myron Schaeffer's backing, other composers and studios became aware of Moog's invention. Only two weeks after visiting the Toronto studio, Moog got a telephone call from Jacqueline Harvey of the Audio Engineering Society, inviting him to attend their annual convention of engineers in New York that October.[14] "We hear you people are doing some interesting things up there." As well as being invited to present a paper, he was offered a free exhibit booth (due to a late cancellation). He rode the bus down to New York to give his first ever academic presentation, "Electronic Music Modules."[15] But it was to the exhibit hall, where about twenty to thirty large mixing consoles, professional tape recorders, and so on were set up, that he was drawn: "I set up a card table—and put these four modules on that we took. On one side of me is this huge tape recorder; on the other side mixing consoles. I was pretty young then. I was thirty—never been to anything like this, never thought of myself as a member of an industry, I was going to make kits after all. I didn't know what the hell I was doing there."

Bob was soon to discover what he was doing there. And he has been doing it the rest of his life: "Alwin Nikolais, the choreographer who does his own scores . . . shows up and then I heard the words, that I later realized were the magic words: 'I'll take one of this, two of this, and that one . . .' We took two or three orders at that show." Nikolais, with the aid of a Guggenheim Fellowship, purchased what would become the first ever commercially made Moog synthesizer.[16] Selling synthesizers was not what Bob Moog had planned. "You know it just happened. It [was like] in these

Figure 6. First sales, AES, New York, 1964

amusement parks where you're sort of going down and you're not quite in control. You know you're not going to get hurt too badly because nobody would let you do that, but you're not quite in control. That's the way it was." For Moog in the fall of 1964 the roller coaster ride of being a synthesizer manufacturer had begun. He had no idea just how thrilling or rough that ride was going to be.

2

Buchla's Box

Some of my best friends are keyboard players.
DON BUCHLA

A T ABOUT THE SAME time that Herb Deutsch was asking Bob Moog if he could make *wooo-wooo-ah-woo-woo* sounds, two experimental composers, Ramon Sender and Morton Subotnick, were collaborating with electronics specialist Don Buchla to develop new devices for making electronic music. Buchla's invention, the Buchla Box, conceived of quite independently from Moog's, could not have been created in a more different place. While the Moog was born in a tiny upstate New York town, the Buchla Box came into existence in radical 1960s San Francisco. The different circumstances surrounding the birth of the two instruments decisively shaped their form and use.

Don Buchla was born in Southgate, California, in 1937, three years after Bob Moog was born. His father was a test pilot in the Air Force and his mother was a teacher. He was raised both on the West Coast (California) and the East Coast (New Jersey). His background appears to have been very similar to Moog's. He is from the same generation, and as a kid he too studied piano.

Buchla has always been involved with music. During his career he has made new instruments, composed electronic music, learned to play flamenco guitar, jammed with the Grateful Dead, and given numerous performances using his own synthesizers and other instruments. He describes himself today as neither an engineer nor a musician but "a traditional builder of musical instruments."

○

On the Edge

Buchla's childhood talent was electronics. Like Moog, he was a tinkerer who got his start building crystal sets and messing around with ham radio. "I had a natural facility for electronics. I never studied it, I just picked it up as a child . . . I sort of regard it as a tool, much as a car mechanic regards a monkey wrench as a tool. It's not something that you study, it's just something that you use. I used it."

In 1955 he enrolled at the University of California, Berkeley, and graduated with a major in physics in 1959. He went on to do graduate work in the field but never finished his PhD. His special talent for electronics was increasingly recognized. The University of California's Lawrence Berkeley National Laboratory had one of the first particle accelerators and was a leading center for the newly emerging discipline of high-energy physics. Buchla worked there building klystrons—devices for producing high-frequency electric fields that accelerate the particles in an accelerator. This turned out to be a good training ground for a career in analog synthesis: "It's partly because I went in there without an engineering degree, and didn't know how anything was supposed to be done." Like Bob Moog, Don Buchla stresses the limitations of academic knowledge. He regards himself as an experimenter; he started off experimenting and he's been experimenting ever since: "I always figured that if I made something that was too popular that I was doing something wrong and that I had better move on . . . I regard myself as more in the avant-garde, kind of experimental phase. And I've always enjoyed *being on the edge*, working on new things, and encountering people that were working with new things."

The 1960s was an opportune time to do new things. The space age was

33

taking off, and electronic sounds had always been part of the mystique of space—the bleeps of the first Sputnik emerging from the background hiss of early radio receivers is etched into the Cold War consciousness. But space meant something else to synthesizer pioneers: it meant a source of employment. NASA needed engineers. Before founding the synthesizer company ARP, Alan Pearlman made equipment for NASA, and Don Buchla's talent for electronics too soon found a new home in space.

Berkeley administered a number of NASA projects. Buchla took part in some of the first investigations of the Van Allen radiation belt in the magnetosphere above the earth's atmosphere. He even directed a project to explore the feasibility of sending chimpanzees to Venus (conclusion: not feasible). He has worked on and off for NASA throughout his career. NASA contracts are one of the few forms of financial security available to the maverick synthesizer designer: "They don't pay much, but they pay more than music . . . and it has been fascinating work."

At NASA Buchla met and worked with the first generation of astronauts. This gave him an opportunity to explore an interest that stretched back to his childhood and that would obsess him the rest of his life: human–machine communication, "just a general interest in how man communicates with machines. I started it as an early child and it continues. And music brings out many of these problems."

◎

The Berkeley Drop Out

The Berkeley Physics Department at this time—during the height of the Cold War—was becoming immersed in politics. Physicists were being asked to make political declarations of loyalty and testify before the House Un-American Activities Committee. The protests against McCarthyism were centered in San Francisco and Berkeley. In September 1964, when UC Chancellor Clark Kerr enforced a gag order against politicking on the

Berkeley campus, the Free Speech Movement was born. Don and his fellow scientists joined the protests. As the sixties unfolded, Don's increasing immersion in the counterculture made a formal career in physics less and less likely. As Don told us, he "dropped out." By this point he had almost certainly "tuned in" and "turned on."

One person who knew Buchla well is Suzanne Ciani. Today, Suzanne is a highly acclaimed Grammy-nominated musician whose New Age–style albums have brought her commercial success.[1] Back in the early seventies, she was a struggling electronic music composer who desperately wanted a Buchla synthesizer and who started to work for Buchla to save up enough money to buy one. Suzanne described Buchla as a "true original thinker, a Renaissance Man," a man with a "vision": "He had glasses and kind of longish hair . . . We called him Buch the spook . . . socially he was pretty shy, and when we were working, in the factory there, Buchla would walk around . . . making these funny noises. He sounded just like a machine . . . He was a very private person, very independent. Very driven, focused and really impassioned by what he was doing—endless energy."

Buchla refused to make concessions to the world around him. He was determined to follow his radical path for the synthesizer, wherever it might lead him.

○

Experiments with Sound

Buchla's overriding talent in all the different areas where he worked was electronics. And with the new form of miniaturization offered by transistors, electronics became a more and more useful tool. Before he developed his synthesizer, several of his early projects involved sound, if not music. With the support of the Indian medical profession, he designed a new form of transistorized hearing aid; at the time, hearing aids in the United States still used vacuum tubes. To aid blind people, he built a device that changed pitch according to its proximity to objects. He also built miniature amplifiers, "but then the FBI got interested and I got uninterested." As Theremin in the Soviet Union found, governments had their own uses for the revolution in electronics.

It was during his time at Berkeley that Buchla became fascinated by electronic music. No doubt it was an opportunity for him to experiment some more. He played around with the standard techniques like *musique concrète*, splicing together a composition made out of the sounds of insects. He owned a one-track tape recorder, but he wanted something better. In 1962 he spotted what he was after at a concert at the San Francisco Tape Music Center. It was a three-track Ampex tape recorder that had been specially designed for film work. He borrowed the machine and started to hang out at the Tape Center.

○

The San Francisco Tape Music Center

The Tape Center had just been formed by electronic music composers Ramon Sender and Morton Subotnick. Sender had come to the United States from Spain as a child refuge from the Spanish Civil War. As a composition student at the San Francisco Conservatory, he had put on his own electronic music concerts.[2] The first one, "Sonics I," in the fall of 1961, had premiered work by the up-and-coming local composers Pauline Oliveros, Terry Riley, and Phil Winsor. Sender encouraged his performers to improvise and called that first concert "bring your own speaker night." It was after this concert that Subotnick, a composer teaching at Mills College, got involved. They were soon putting on electronic music concerts together. Their 1962 series ended with what Sender describes as an early "happening": "We projected a film on the patio concrete and had the dancers dancing in it. And we had one person inside a piano . . . the audience was encouraged to walk around the building . . . I found an old cast iron aluminum laundry machine which we filled full of rocks and turned on, and had a long extension cord that allowed us to wheel it down the corridors. It was quite thrilling . . . and sort of strange." Sender had introduced more visual elements and audience participation because he discovered that audiences did not like just listening to tapes. This was a problem endemic to electronic music before the synthesizer: without a performer, a concert was terminally boring to watch.

Subotnick had some equipment in a garage, and Sender came up with the idea of asking other composer friends, such as Oliveros, to pool their equipment in one studio location. Thus was the San Francisco Tape Center born, first located in a house on Russian Hill and then on Divisadero Street.

The center soon acquired the reputation for staging some of the most avant-garde music and multimedia happenings in the Bay area. Composers John Cage and David Tudor gave a much-celebrated series of concerts. Karlheinz Stockhausen lectured. Terry Riley and Steve Reich were just starting their careers. Terry Riley's "In C," commissioned by the Tape Center, was premiered there in 1964, and Steve Reich's "It's Gonna Rain," also commissioned by the Tape Center, was played in February 1965.[3] Nearly every concert involved some form of improvisation.

The Tape Center was also a working studio. It made electronic music in the traditional way, by cutting and splicing tape. The musicians used just about anything they could lay their hands on for sounds. They raided junkyards, borrowed from local industries and universities, and accumulated piles of war surplus. As Bill Maginnis, the center's technician, told us: "There were bits and pieces of World War II bombers that we had. Also . . . someone had raided the rad lab [Radiation Laboratory] at Berkeley, and we had bits and pieces of stuff that said 'University of California Department of Physics and Cosmic Rays' on it."

Subotnick and Sender were, however, searching for better ways to do things. They wanted to "move away from cutting and splicing to get something that was more like an analog computer."[4] What they were thinking of was a "black box" for composing. Instead of having to go to a big studio, composers would have access to this inexpensive piece of equipment in their own homes. Sender and Subotnick on the West Coast were driven by precisely the same needs as Deutsch on the East Coast. They too wanted to use the new electronics to help electronic music making become more portable and affordable.

Bill Maginnis still remembers the moment he was hired as the center's technician. He was a working jazz drummer and had gone there to get a circuit diagram for a ring modulator. Ring modulators are devices first used in radio for multiplying two waveforms together. The resulting output (the sum and the difference between the two frequencies) is musically interesting because the output frequencies have no musical relationship to the input frequencies. They produce very strange bell-like sounds because of the altered harmonic structure. Maginnis: "So I walked into the Tape Center and Ramon Sender was standing there . . . He said, 'If you had a schematic, could you build it?' And I said, 'Sure, yeah.' He reaches in his pocket . . . starts fumbling with his keys, he says, 'Here's the key to the front door, here's the key to the lab.' I said, 'What, what, what?' He said, 'Oh, you're our new technician.'"

Maginnis first became aware of Buchla and his talent for electronics when Buchla offered to help him build the ring modulator: "You don't have to go through all this trouble, it's easy, just give me four diodes and I'll build you a ring modulator." Maginnis still has that hand-soldered ring modulator, which was the prototype for the one Buchla built into his first synthesizer.

Buchla saw what the Tape Center was doing and realized that the field was ripe for the application of electronics to music. He started to discuss his ideas with Subotnick. Maginnis too had a role to play; he found he was translating between the engineers and the musicians.

The idea of an "intentional" electronic music device—one that was designed from the start to make electronic music rather than using equipment designed for other purposes—slowly developed. Subotnick suggested that using a light source to control sound might be promising.[5] Despite being skeptical of this approach, Buchla built the device, and less than one week later showed up with it on a plywood board. Subotnick: "This was a disk that rotated and had little holes in it, with a light that shone through it onto a photocell. By changing the position of the holes, you could change the harmonic structure of the signal at the output of the photocell."[6]

Buchla's skepticism proved to be correct. After building this device, he announced to Sender and Subotnick: "This is the wrong way to do it."[7] Maginnis too was experimenting with light and photoresistors, which he used in conjunction with relays to make a rudimentary form of envelope generator. Later he turned this into an electromechanical sequencer by introducing step relays and a dial. It only had ten even steps before needing to be reset, so it was not useful for making any sort of rhythm.

The funding of the Tape Center was always precarious, but finally Sender and Subotnick persuaded the Rockefeller Foundation to give the center a small grant, $30,000, to run their program for the 1964–65 season. Five hundred dollars was allocated to Buchla to work on building the intentional electronic music device.

Buchla's familiarity with silicon transistors and his knowledge of analog computers (from working in physics) led him to voltage control: "I had this idea you could take voltages and multiply and mix them and things like that. So it wasn't too far a cry for me to attach a voltage to the pitch of an oscillator or to the amplitude of a VCA . . . But as soon as I added voltage control to the elements of the synthesizer it became a different ball game because you could parametize everything. You weren't limited by how fast you could turn a knob to get between two states of a parameter."

Buchla, like Moog, realized that voltage control had the potential to revolutionize the way sounds were controlled. But Buchla was after something different; he wanted to eliminate the time-consuming practice of splicing tape altogether. By thinking of ways to do this electronically, Buchla was led to the electronic sequencer—a device that later was used to make much influential pop, rock, and dance music. A sequencer produces predetermined control voltages in a cycle or sequence and can endlessly recycle these control voltages at a frequency predetermined by the user. In Buchla's first sequencers there were three control voltages per step, and his first "sequential voltage sources" had eight voltage steps (later sixteen). It was a rudimentary way to program a voltage-controlled oscillator to play a series of different pitches one after the other—in effect, a tune. As the notes repeat, the sound can be shaped and subtle changes in rhythm and timbre introduced. Buchla: "You wouldn't have to splice 16 pieces of tape together if you wanted a sequence of 16 notes. You could simply take my sequencer and set the time and the pitch for each interval. So that required, of course, a voltage-controlled oscillator and sequencer, and from then it led to a bunch of other ideas."

The idea of using electromechanical ways to sequence a series of sounds was very much in the air. Peter Zinovieff in London was also working on rudimentary sequencing devices, as was the pioneering jazz musician and inventor Raymond Scott.[8] But there was no doubt that Buchla, with his electronic way of doing it, had achieved a breakthrough. The sequencer was the one module that Moog eventually "copped" from Buchla.[9] Sender and Maginnis were impressed by what Buchla was developing. Maginnis: "Ramon had seen what Don was doing and came back and said, 'Hey, he's got a thing sort of like your dial that is really, it's incredible.' I said, 'Really?' And what it was was a sequencer."

○

"An Exciting Day"

Bill Maginnis still remembers the day that Don brought his prototype Buchla Box into the Tape Center for the first time in late 1965: "And he says, 'Well, here it is.' . . . Well, everybody was gathered around up until about 10 o'clock in the evening playing with it. They all left and I stayed

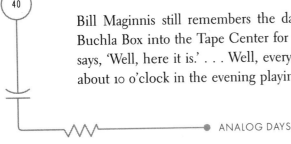

40

there all night long, playing with this thing." Before leaving in the early morning Maginnis programmed the sequencer to play the first eight notes of "Yankee Doodle." "I knew Pauline [Oliveros] was coming in in the morning and she wanted to work on something . . . So I get this call at 10 o'clock in the morning from Pauline saying, 'How do I turn this damn thing off?'" Pauline had a very different way of making electronic music and was rather indifferent to Buchla's device.[10] We asked Buchla about that day: "It was an exciting day . . . and I hooked up the sequencer to the voltage controlled oscillator and turned the knobs. And Terry Riley said, 'I think I'll do some music on that.'"[11]

Maginnis was amused to point out to us that the original Buchla Box, now at Mills College, still has modules that are tack-soldered with components attached to posts. That first prototype also had gaps for modules that Buchla had yet to design. Like Moog, Buchla had no idea how important synthesizers would become. At the Tape Center, it wasn't as if the whole future of electronic music-making was about to change. Maginnis: "It was incomplete . . . It was just another device going in. But this one said San Francisco Tape Music Center on it instead of Hewlett Packard."

Buchla was reluctant to call his device a synthesizer. For him the word synthesizer had (and still has) connotations of imitation, as in the word "synthetic," meaning rayon and other man-made fibers. He did not regard his new instrument as a vehicle to imitate or emulate the sounds of other instruments. He eventually called his instrument the Buchla Music Box Series 100. It was often referred to as the Buchla Box or simply the Buchla.

Buchla came up with his invention without any knowledge of what Moog was doing on the East Coast at about the same time. Buchla got started a few months ahead of Moog, but Moog's first prototype was finished in the summer of 1964; Buchla's appeared in the fall of 1965. The two inventions were very similar: both were modular, were connected by patch wires in flexible combinations, and used voltage control for the oscillators and amplifiers.

BUCHLA'S BOX

Figure 7. Buchla 100 Music Box

There were, however, some differences. They initially designed different modules: Buchla had a ring modulator and sequencer; Moog later added a sequencer and manufactured a rack-mountable ring modulator designed by Harald Bode (designed before either Moog or Buchla made their prototypes). On the other hand, Moog had a filter, which Buchla later added. There was also the one difference that would become important for the future of the synthesizer: Buchla's lack of a standard keyboard. He provided instead a series of touch-sensitive pads that produced three different voltages the harder you pressed the pads. On the first Buchla Box he built for the Tape Center, these pads were used to control a set of *musique concrète* tape loops that Subotnick and Sender had made.

Moog and Buchla may have started at the same point, but they ended up at very different places. They were both socially awkward young men who found solace in their hobby of electronics. Both had some familiarity and interest in music and did their graduate work in physics departments at leading American research universities. Both were interested in exploring new things, and neither of them had any experience of the commercial world of instrument manufacture. Their hobbyist background and independence meant that they could tinker and experiment freely without the constraints that a corporate setting might have imposed. The hobbyist tradition provided them with a community and a network within which they could comfortably operate and design new devices. Both had enough training in, and understanding of, electronics to recognize the importance of the development of transistors and advances in voltage control. Crucially, both of them, at key moments, encountered the world of avant-garde electronic music—Moog in New York and Buchla in San Francisco. But the degree to which they were immersed in the world of the avant-garde would produce important differences in their synthesizers.

◯

Strange Arrays

Some people have said that Buchla is anti-keyboard, but they misunderstand him. He considers the keyboard a perfectly good way of doing what it does well, which is making polyphonic music based on a twelve-tone chromatic scale. It just never occurred to him that such a device was an appropriate way to control electronic sound: "It was a potentially new source and therefore instruments based on it would probably be new and different." Buchla was committed to this vision of doing something completely new.

He felt that going the keyboard route was reverting to an older technology: "I saw no reason to borrow from a keyboard, which is a device invented to throw hammers at strings, later on operating switches for electronic organs and so on."

Buchla wanted something more imaginative as a controller, something that would connect the performer to this new source of sound. No doubt his life-long interest in human–machine communication was part of his motivation to improve on the keyboard. His arrays of touch-sensitive metal pads housed in wooden boxes were designed to be more user-friendly. Buchla had a special name for these devices: kinesthetic input ports. Although these pads could be tuned to play the twelve-note chromatic scale, Buchla's whole design philosophy was to get away from the constraints of the standard keyboard.

His attitude was shaped by the experimental composers he met at the Tape Center. John Cage and his collaborator, David Tudor, were exactly the sort of artists with whom Buchla identified. Cage used Buchla's touch pads to control one of his favorite pieces of equipment, the voltage-controlled FM radio receiver (which he used as a source of electronic sound for musical performance). Each pad was used to control a different station. Buchla's first-ever sale was to David Tudor, for whom he designed a set of five circular pads that could move sound around a space, from speaker to speaker.

In Buchla's vision of a keyboardless synthesizer, the operator would be stimulated to explore the new sounds of which the new instrument was capable: "A keyboard is dictatorial. When you've got a black and white keyboard there it's hard to play anything but keyboard music. And when's there not a black and white keyboard you get into the knobs and the wires and the interconnections and the timbres, and you get involved in many other aspects of the music, and it's a far more experimental way. It's appealing to fewer people but it's more exciting."

Mass appeal was not Buchla's goal. He saw no particular need to standardize his modules in terms of the conventional musical scale. In other

44

words, he did not share Moog's commitment to the one-volt-per-octave standard. The Buchla, built without a conventional keyboard, could not be described according to this standard at all. Buchla had no need to use exponential converters on his oscillators. "Our original oscillators were actually linear rather than exponential. In fact the original ones were not even linear; they were something in between. So I can't give a volts-per-octave figure."[12]

○

Buchla Goes Bananas

Another subtle difference between the technologies of Buchla and Moog lay in the sorts of patch cords they used. Patch cords are the wires that allow the operator to flexibly connect up the different modules on a synthesizer. Buchla used two different sets of patch cords with different sorts of wires and plugs for "signal" and "control" voltages.[13] Signal voltages are voltages in the audio range of frequencies, while control voltages are typically of lower frequency. Buchla routed the control voltages through unscreened wires with plugs at each end, known as banana plugs, that could be stacked one on top of the other. The signal voltages were routed through screened wires and phono plugs. "It was to prevent the user from interchanging the two. Because a module that optimized for one function was definitely not optimal for the other function."

Moog, on the other hand, used standard jack plugs and screened wires for all his patch cords, making no distinction between the two sorts of voltage. Buchla felt that his separation of signal from control voltages made more sense electronically. For example, he could invert control voltages (turn a positive voltage to a negative one), but it made no sense to invert audio voltages. Modules designed for the audio range such as voltage-controlled amplifiers (VCAs) could be differentiated from ring modulators which worked in the control realm. Mixing the two together was, for Buchla, like building a "car which could also float." In short, a disaster.

45

Another advantage Buchla claimed for keeping control voltages separate from signal voltages was that it helps a composer to understand more easily the structure of a complex patch. "You can look at the system and see where it's at. That invites more complex systems. Mort Subotnick was I think 'King of the Wires' on that . . . you could put in a lot more complex structures and still not confuse them with the sounds." In other words, by stacking the banana plugs one on top of another, a complex pattern of control voltages could be established without inadvertently confusing this structure with the audio sounds themselves.

This difference was much less salient for Moog: "In order to separate them you'd have to think you would never want to use an audio signal as a control . . . and I just never saw that—that's not something that we'd want to decide up front. I'm not sure there is an advantage to doing it his way, other than a conceptual [one], there's less chance to mix up control voltages with the audio, so maybe his was easier to use, to think about for musicians, but mine was more versatile."

Moog's rather different design philosophy can be seen at work here. Why constrain the users? If the users wanted to mix up control voltages with audio, why not let them do so? The Moog way of doing things was to let users decide these sorts of questions for themselves. On the other hand, from Buchla's perspective he was providing composers of experimental music a way to make very complex patches easier to follow and thereby encouraging them to experiment.

◎

The Source of Uncertainty

Buchla created his own unique series of names for some of his modules. All synthesizers, including Moog's, have devices for generating noise that are particularly useful for producing percussive effects and other sounds like rain, waves, thunder, and wind. White noise, as it is known, is analogous to white light in that it contains all sound frequencies. Because human hearing gives prominence to higher frequencies, we hear these slightly more in white noise—to us, it sounds like steam escaping from a radiator. Pink noise is white noise that has been filtered to boost the lower frequencies (formally, it has equal energy per octave). Rather than call his noise sources white noise or pink noise, which became the terms favored by most manufacturers, Buchla on his System 200 (introduced in 1969) named them "the source of uncertainty."[14] The source of uncertainty served primarily as a random source of control voltages (which could involve, for instance, selectable probability distributions and voltage-controlled rates of change) and thus was a way of introducing randomness into compositions.

Another module on the System 200 was called the "multiple arbitrary function generator." Buchla maintains that these names, like "kinesthetic input port," were chosen because they described as accurately as possible what the device or module did: "It wasn't that I was trying to be different; I was trying to be accurate . . . 'Source of uncertainty' meant precisely what it says. 'Random function generator' means randomness. Randomness suggests just that, a random walk or something. 'Source of uncertainty' means uncertainty is a thing that you deal with. It's not something random. If you want some uncertainty it's a positive attitude." Buchla was asserting the rather different esthetic behind his vision.

The Buchla Box was designed for musicians who wanted to produce a complex piece of music in real time. The sequencers and the source of uncertainty enabled the operator to set off complex chains of events that could feed back on each other. It favored a certain sort of composer and style of composition. Moog: "There are also people like Morton Subotnick and Suzanne Ciani who are concerned, as Cage was, with production of music as process, where to realize your music, you would organize a very complex system. The Buchla modular system was designed with this sort of composer in mind more than ours was. It has a lot of capability for triggering sources in sequence, for turning on and off different sources, and for creating a very complex organization of a modular system. You can literally set up a machine that will produce an interesting sounding piece of music by itself."[15]

Figure 8. Morton Subotnick with Buchla 100, 1969

The acknowledged "king of the wires," Morton Subotnick, found the Buchla to be a totally new way of composing electronic music. This is how he made his acclaimed album *Silver Apples of the Moon* (1967; title from a poem by W. B. Yeats), created on the Buchla 100.[16] Subotnick: "I purposefully did not know what results I was after. I believed that with this new instrument, we were in a new period for composition, that the composer had the potential for being a studio artist, being composer, performer, and audience all at once, conceiving the idea, creating and performing the idea, and then stepping back and being critical of the results."[17]

Suzanne Ciani has used Moogs and Buchlas. Although players obtain immediate aural feedback from both instruments, and there is tactile satisfaction from interacting with keyboards, touch plates, and potentiometers, Buchla was interested in involving as many senses as possible and added multiple arrays of indicator lights on his System 200: "There were lights on everything, because that was your feedback . . . even [on] your source of uncertainty. If you wanted to know the rate or the intensity of that voltage, the light told you that, so you could tell, 'Oh, gee, I'm having a very rapid rate of uncertainty,' or a slow rate."

Ciani felt that the Buchla offered more subtle controls such as the three-dimension sequencing ability of the multiple arbitrary function generator and the ability to mix oscillator waveforms (a feature the VCS3 would also have).

○

Art for Art's Sake

Buchla was developing a unique way of doing things. His synthesizer was geared toward a certain sort of user. And there were like-minded composers who saw the merits of his approach. No less a figure than Vladimir Ussachevsky visited the Tape Center and ordered three of Buchla's systems, one for each of his identically furbished studios at Columbia-Princeton. Buchla's synthesizers appealed to Ussachevsky's way of working as a composer: "The Buchla synthesizer was logically arranged to be more accessible to composers' thinking . . . than Moog's synthesizer. This is not to say that Moog's modules were not awfully well made, very enduring, and in certain cases certain of his modules *at that time* were superior to Buchla's in quality . . . I felt that somehow it did not have enough flexibility."

Ussachevsky's order led Buchla to form his own company, Buchla and Associates. His first shop on Ashby Street in Berkeley was a storefront, little

49

bigger than a garage: "I wouldn't call it a factory . . . we would lay our equipment out on the sidewalk to have enough space to work with things and on rainy days we would just [acts out picking up the gear]." Later, he had the studio shop in Oakland where Suzanne Ciani first worked for him. Most of his early sales, before word spread, were for customized modules ordered by composers who visited the Tape Center.

Buchla has always been uneasy with commercial production. He regards himself as an artist who, as Ciani put it, "lived off his art." Buchla: "I don't regard it as important to be a commercial success . . . It's a fine line to tread, and you want to stay in the arts at the creative end of it and not mass produce anything, and yet you still have to make a living if you choose to do it."

His artistic vision has had consequences, however, for the commercial success of his synthesizer. Because he did not set himself up as a mass manufacturer of synthesizers, he was never able to sell many systems. Also, being a small designer based in one location meant synthesizers had to be sent back to his studio for repair and service, and all of this trucking around occasionally damaged their delicate electronic innards.

○

Buchla and CBS

In 1969 Buchla did venture into a more commercial form of production when he licensed his designs to CBS. Electronic music was taking off, and CBS hired staff who developed slick demonstrations of the Buchla. For a while the Moog company was worried by the impact that Buchla/CBS might have, but CBS's arrangement with Buchla fell through.[18] He does not like to talk much about this episode, which was not a happy experience for him.[19] "They thought they wanted to branch into musical instruments. There were quite a few CBS systems sold. They really got into production. But then they dropped the ball. It was too small a market and too narrow a focus. Their Fender guitar business was booming and they wanted the larger market that that represented, and it was clear to them that electronic instruments were destined to be on the fringe."

When Ciani first met Buchla, he was still bitter about the CBS experience: "He was very alternative, so he knew that he was making something that was, in a sense, anti-establishment. It was new. The establishment, as we called it at that time, was dangerous because they took things over and they homogenized them." It must have been difficult, indeed, for Buchla as a countercultural artist to be dealing with part of the "establishment" like CBS.

With the CBS episode behind him, Buchla returned to his low-key operation, making systems for his own use and for a small number of like-minded composers and academic studios. It's a way of operating that he has followed to this day. "It's a niche market . . . The stuff got popular because it sold to a lot of education institutions and a lot of them had high exposure. I mean until this day . . . if I go into a store and buy something and they see my name on my credit card and say, 'Boy, I learned on your system, it's in such and such a school.' Twenty-five years ago it was very common." One university to buy a Buchla was Harvard. The famous minimalist composer John Adams learned synthesizer on a system that the Music Department purchased in 1969. Although less well-known than the Moog, Buchla's synthesizers have been a source of inspiration to many artists and composers.

Buchla never was part of the electronic instrument manufacturing fraternity. He did attend one AES meeting, in 1971, where he met Bob Moog for the first time. But while there he "offended all kinds of people." By refusing to rent a booth and instead playing a concert, and thus gaining lots of free publicity, he broke most of the rules for such conventions. And to this day Buchla continues to break all the rules. Over the years he has made an extraordinary variety of analog and digital synthesizers and different sorts of controllers.[20] His synthesizers are still known only within a small circle of people, but he has kept ownership of his own company and after thirty years is still making synthesizers and practicing his art.

Moog, in 1975, perhaps chastened by his own experiences in the commercial world, came to appreciate Buchla's anticommercial stance: "I have to admire what Don Buchla has done. He hasn't allowed himself to limit the complexity of his instruments to meet the demands of the so-called marketplace. It has been the conventional wisdom for some time that a complicated piece of electronic music equipment can't be sold off the music store floor. Buchla has chosen not to worry about this."[21]

Buchla has garnered the reputation of being the odd ball in the synthesizer field. What counts as odd, of course, depends on what is going to count as normal. Back in 1965 it was not at all clear that the synthesizer would appeal to a mass market, and also it was not clear that it would become primarily a keyboard instrument. To understand how all this occurred, we return to Trumansburg, to see what Bob Moog was up to.

3

Shaping the Synthesizer

> The Moog filter design is as unique to the sounds of the synthesizer
> as the Steinway steel frame is to the piano.
> Herb Deutsch

Bob Moog was excited. He may have been only a graduate student, but he had made a start in the electronic music instrument business. He had built his first synthesizer modules, he had three precious orders, and he had his beloved shop. And the shop was soon going to expand.

The Modular Moog synthesizer, which was advertised for the first time as a "synthesizer" in 1966, emerged over a period of time from a thousand different design decisions and a thousand conversations. It was an *innovation* rather than an invention.

The history of technology tells us that inventions are two a penny. There are many, many people who invent new things: machines, processes, tools, gizmos, gadgets, widgets, and the like. Such people are often portrayed as unsung heroes, ahead of their field, unrecognized in their own time. But singing in the bath is not the same as singing on stage. There are very few people who successfully turn their inventions into a real product that can be manufactured, marketed, and sold. The field of electronic music instruments is littered with inventions, but there have been very few true innovations.

One of the most spectacular inventions is Thaddeus Cahill's Telharmonium, developed in 1901.[1] This enormous instrument, which weighed 200 tons, generated sound from giant alternators. It was installed in Telharmonium Hall in downtown Manhattan, and its sound was piped to nearby hotels and restaurants. Technical problems of crosstalk with the ordinary telephone cables and a rather annoying timbre led to its failure. It was, however, a precursor to one of the few true musical innovations—the immensely popular Hammond organ invented by Laurens Hammond in 1933.[2] Hammond used a series of spinning tone wheels (a miniaturized form of the Telharmonium sound source) and took advantage of tube amplifiers and loud speakers (unavailable to Cahill).

Another notable inventor was Friedrich Trautwein, who in 1928 developed the trautonium (which Goebbels was keen to use for Nazi propaganda) and who, with composer Oskar Sala, went on to produce the mixturtrautonium used for the sound effects in Hitchcock's *The Birds* (1963).[3] Maurice Martentot developed the Ondes Martenot in 1928; it was used in classical performance and had a special repertoire written for it, including works by Varese, Messiaen, Boulez, and Ravel. To this day it is still used in France. Luigi Russolo, a member of the Italian futurists, in 1914 developed an early mechanical form of the synthesizer. His *intonarumori* (noise instruments) provoked scandal but remained as oddities.[4] However, like the theremin and the Ondes Martenot, Russolo's *intonarumori* found a home in movies.

○

The Listening Strategy

Bob's success as an innovator can be traced to one key factor: he listened to what his customers wanted and responded to their needs. Rather than telling them "this is the way things are going to be," he devised a strategy over the years for learning how other people wanted things to be. He learned from going on the road, entering their homes and studios, and bringing them to Trumansburg, first for a summer workshop and later to his own factory studio. He saw more clearly than anyone that his own fate as a manufacturer was tied to the success of the field as a whole, and he devised ways of nurturing that field, such as by starting his own magazine for electronic music.

Moog's strategy does not appear to have been deliberate. It was not that on the bus ride home from the 1964 Audio Engineering Society meeting Bob planned out the next three years. As he constantly reminded us, "I didn't know what the hell I was doing." He fell into what he was doing, but he learned as he fell. He learned from what he found happening to him, around him, and in the culture. And above all it was fun. Bob had a blast during those early years. What he enjoyed doing most we know already— fiddling, diddling, and futzing with electronics. But he soon discovered he enjoyed something else—meeting his customers, many of whom became his life-long friends. Moog: "Nikolais and Siday and Hillar and all the people I did business with in the early days have remained collaborators and friends and customers throughout the years. I've gotten to know them all . . . They've been very valuable to me both as personal friendships and as guidance in refining synthesizer components."

That guidance began with his first three customers. Alwin Nikolais, Lejaren Hillar, and Eric Siday represented a spectrum of needs. Nikolais wanted a bunch of modules on which to make avant-garde music for the dance troupe he choreographed. He had previously used tape loops. Hillar was an academic composer who headed the well-known University of Illinois electronic music studio, and Siday was a commercial musician. Moog formed a particularly close relationship with Siday, whose order turned out to be important for the future of the synthesizer. Siday wanted not just modules but a complete system, and he had the money to pay for it.

○

Eric Siday

Eric Siday was one of the best-paid commercial musicians of the day. Trained as a violinist at the London Royal Academy of Music in the 1920s, he played in silent films before moving to the United States in 1938. When radio and TV came along, he helped invent a new occupation: creating electronic jingles, sound signatures, and sound logos that, in five seconds, identify a product or corporation on TV or radio. He used oscillators, tape loops, splicing, and any other technique he could lay his hands on to ply his new trade. One of his best-known signatures was the burps of coffee percolating in a Maxwell House coffee commercial. The little electronic ditty that introduced every CBS television show as an announcer said "CBS presents this Program in Color" was heard by millions, and it earned him $5,000 for each second of sound.[5]

After Moog's AES presentation, Siday and his technician came to Trumansburg. Moog: "We sat down and conceived on paper a whole modular system that was going to cost him $1,400. It was an incredible load of stuff . . . something like ten or twelve modules." As far as Bob can recall, "this is the first time when a system the size of a synthesizer was actually talked about between me and a central customer." Bob had already designed many of the modules Siday wanted. He had voltage-controlled oscillators, amplifiers, and filters, and he had envelope generators. But now he had to put them all together into one system. He had to design a workable keyboard and a cabinet in which to house everything. He also had to come up with a price. "I knew about what the material would cost. I thought I had an idea of how much work would go into it, once we knew how to build them. Well, I didn't know! A lot of time it was just feeling and guesswork. I was not a businessman at the time . . . I literally didn't know what a balance sheet was." The business side of things was something Moog never enjoyed or mastered. "Feeling and guesswork," as he was to discover the hard way, didn't keep a business running.

After six months' work the system was ready for delivery. Siday's studio was in his home, a grand ten-room apartment in an Upper West Side building, the Apthorp. Siday had taken over the elegant living room and maid's bedroom for his work. "It was completely filled up with instruments and half-instruments and stuff taken apart and stands and whatnot." It was to this home studio that Bob was to deliver his first complete system. There was just one problem: How was he going to get this rather large synthesizer down to New York?

Bob did what he always did—he rode the bus: "We worked months and months and we sweated it out. First product, and we were very proud of it. My coworkers and I packed it up into some cardboard cartons and I took it on a bus—a big synthesizer—it takes two men to carry each box . . . I got it into a taxi after the bus and got it to the Apthorp." It was now morning and Bob was a "little bit strung out" after his night on the bus:

> It's eight o'clock in the morning and Edith [Eric's wife] is watching this like a shy child . . . from one of the doorways. I put one of the boxes down and take the lid off and take all the packing out and spread it on the hall floor. Take one of the instruments out and set it up. Begin to unpack the next box and I didn't realize this, but Edith is slowly losing control of herself. She's watching this. I was busy unpacking. I'm setting the second box up . . . All of a sudden she loses control of herself completely. She screams out, 'Eric, more shit in this house. All you ever do is bring shit in this house. One piece of shit after the other!' . . . and somehow I got the instrument set up and [got] out of there that day.

Over the years Moog designed many customized modules for Siday, including a keyboard where each note was individually tunable, although he had trouble delivering them on schedule. On another famous occasion, Edith (who by now had become close with Shirleigh Moog) bawled Bob out for keeping Eric waiting for a piece of equipment. Bob found it hard to make a deadline—he was nearly always late.

Other customers were important for the future of the synthesizer, and none more so than Wendy (formerly Walter) Carlos, who pushed Bob to perfect the technology. It was at her urging that Bob designed his first touch-sensitive keyboard. She also came up with the idea of adding the portamento control and the fixed filter bank (a form of graphic equalizer), which eventually became standard features. The portamento control,

57

Figure 9. Eric Siday with Moog synthesizer in home studio, 1965

which allows the voltage generated by one key to slide smoothly to the next, was particularly important for live performance. It was rock performer Keith Emerson's favorite feature of the Moog.

◎

To Key or Not to Key?

Listening to what customers wanted was all very well, but what if customers disagreed over what they wanted? Moog soon faced this problem. In 1965

Vladimir Ussachevsky contacted Moog and expressed interest in buying some of his modules, but in the end he bought very few. Instead, he bought three complete systems from Don Buchla, which, unlike Moog's synthesizers, had no keyboard. Ussachevsky was well-known for opposing the keyboard. The standard keyboard just did not fit with his conception of how electronic music should be made. And Ussachevsky was not someone Bob could easily ignore. Ussachevsky had started off as an engineer and switched to music later. He spoke the language of engineering, and the way to Bob's heart was, of course, through circuitry.

Soon Ussachevsky and Bob were discussing one specific circuit—the envelope generator. Normally such a module is used with a keyboard as a trigger to make the sound of, say, a plucked string. Ussachevsky wanted an envelope generator, but he didn't want it to be triggered by a keyboard. Instead, he planned to use it with tapes to shape recorded sounds. Moog designed him a special module with the envelope triggered by an external switch rather than by a keyboard.

Moog's interaction with Ussachevsky was important not only because it conveyed to him the depth of hostility toward the keyboard that existed among some serious composers but also because it led to the standard way to describe the main functions of an envelope generator in terms of T_1 (attack time), T_2 (initial decay time), E_{SUS} (sustain level), and T_3 (final decay time). Eventually ARP simplified this to the ADSR (attack, decay, sustain, and release) nomenclature still in current usage.

That Ussachevsky had purchased his rival's synthesizers and was opposed to the keyboard gave Moog food for thought. Moog was not strongly wedded to the keyboard as a controller; it was just one option. After all, he built theremins, and these were about as far removed as you could get from a keyboard-controlled instrument. But not everyone agreed with Ussachevsky. One composer who saw things rather differently was Herb Deutsch, who had been Moog's first collaborator: "Bob didn't want to have a keyboard, because he had talked to Vladimir Ussachevsky, and Vladimir said, 'Oh, no, you don't want a keyboard, because then people are going to think of it more traditionally. You'll be, it'll be dominated by the need to be tonal, or at least relate to tonal design' . . . And I simply thought, Well, it didn't bother Schoenberg to have a keyboard. I mean, he still created atonal music and you still have the freedom to do anything you want. And I persuaded Bob to do a keyboard on it." Deutsch felt, in other words, that a serious composer would be able to overcome any conventional associations the keyboard evoked.

The commercial appeal of the keyboard was something Deutsch and Moog both recognized as well. Walter Sear also put pressure on Moog to stick with the keyboard: "Buchla's [synthesizer] had nothing to do with musicians. I kept saying without a keyboard it's not anything that a musician [could use], you know, all musicians, in those days, had to have some background in keyboard." And commercial musicians were voting with their feet—or rather their hands. Eric Siday had wanted keyboards. So too did other commercial musicians.

For many musicians, the simpler-to-operate Moog with its keyboard interface was more appealing. Moog, no doubt impressed by what most of his customers wanted, decided to stay with the keyboard, himself regarding it as a "general controller" with which to adjust a variety of modules on the synthesizer and not simply as a way to play melodic music.

◎

Keyboard Culture

As the Moog synthesizer evolved, it increasingly became seen as a keyboard instrument. Many of the photographs of the Moog (in the media, on record albums, and in promotional literature) show the keyboards prominently displayed. In these early pictures the right hand of the operator is usually on the keyboard while the left hand is outstretched adjusting the knobs. We asked Bob about these pictures: "The keyboards were always there, and whenever someone wanted to take a picture, for some reason or other it looks good if you're playing a keyboard. People understand that then you're making music. You know [without it] you could be tuning in Russia! This pose here [acts out the pose of the left arm extended] graphically ties in the music and the technology. So there are probably a zillion pictures like that."

The need to show that "you're making music" was something of which Bob was all too well aware. The modular Moog synthesizer is a very odd musical instrument, because, unlike most instruments, you cannot immediately get a sound out of it. It first has to be plugged in, patched up, and connected to an amplifier and loudspeaker. On first encountering the synthesizer, people often expect to hear a tune. Jon Weiss discovered this when he brought the synthesizer on the set for Mick Jagger's use in the movie *Performance*: "I had to go through this with the English workers, saying 'Agh it's a fabulous sanitizer and what does it do?' You know, 'Play us a tune' . . . Moog heard that so much that in one series of synthesizers he put a little speaker and amplifier in one so that you could actually hear something. People couldn't conceive that this is an instrument but it doesn't do anything."

A keyboard is immediately recognizable; it's an icon and "looks good" and invites people to come and play it. And it is here that the wider culture and particularly the dominance of the piano played a role in shaping the synthesizer. Over time, almost inexorably, the Moog synthesizer became a keyboard synthesizer. By the time the Minimoog was developed in 1970, it was, de facto, a keyboard instrument.

Moog has eventually come to see the wisdom of not following the strict approach of Ussachevsky and Milton Babbitt, another giant of electronic music composition who co-directed the Columbia-Princeton studio in Princeton and was closely associated with composing on the RCA synthesizer. Moog: "Ussachevsky and Babbitt have always talked as if they were the light and the way and everyone else should follow them. But actually their concept is very narrow, I think. They have dictated that certain things shall not be. There shall not be keyboards and that sort of thing."

With hindsight, we can see that Bob's decision to design his system around a keyboard was a propitious one. But again, hindsight should not mislead. At the time, this was just one of many little decisions being made. It was not that Moog looked into his crystal ball and foresaw that the synthesizer would become a keyboard instrument. It was just that if musicians wanted keyboards, who was he to stop them having what they wanted? His strategy of listening and responding to customers meant that the synthesizer as a keyboard instrument was being shaped by the wider culture. And that culture would, in turn, be shaped and changed by the Moog.

○

The Ribbon Controller

Although Moog went down the keyboard path, he also developed other sorts of interfaces such as the ribbon controller (also known as the linear controller or stringer).[6] This two-foot-long narrow rectangular box has a taut gold-plated metal band strung over a resistance strip running along its length. By moving a finger along the band and pushing down the musician can vary the control voltage smoothly in direct proportion to the point of contact. The ribbon can be used to control the pitch of an oscillator in a similar way to moving a finger up and down a violin string. It can also be used for vibrato, or to control other modules like the filter.

Musicians came to value the ribbon, especially in live performance. One such musician was Chris Swansen, a Trumansburg-based jazz musician. By rapidly sliding his finger backward and forward along the controller, Swansen produced the same effect as a trombonist sliding the arm of his trombone in and out. Swansen played the ribbon with one hand while he played keyboards with the other. It was also useful in the studio. Malcolm Cecil and Bob Margouleff used the ribbon controller with Stevie Wonder to produce the classic "burump" of the bass on "Boogie on, Reggae Woman," and Paul McCartney used it on the Beatles track "Maxwell's Silver Hammer."

Rock keyboardists took to the controller with a vengeance for live performance. For Keith Emerson, it was an indispensable part of his stage act.

The shape of the ribbon controller evokes the guitar, and Keith Emerson wielded it like an axe. Videos of Emerson, Lake and Palmer in performance show Keith standing on stage in all the pomp rock regalia of the day lifting his ribbon controller upward from his groin as the music swells and climaxes. Keith even had toy rocket motors attached to the ribbon controller to fire up at such moments for added pyrotechnic effect (once in rehearsal he was demonstrating the rockets to a fire marshal and was lucky to have had only his thumbnail blown off).

But in the middle of all these pyrotechnics Keith stumbled upon a new use for his ribbon controller: "The ribbon got so worn down, and I was almost complaining, 'Oh God, I've worn this out. I should really call Bob up and get a new ribbon.' But then I realized that by short-circuiting my thumb across the ribbon and the actual bar it made these machine-gun sounds. I thought, 'No, this is great!' There again, it was just, the instrument wore out, but I got other sounds out of it as a consequence." Even in the Moog's imperfections musicians could often find a use.

Running around with a big electronic phallus in live performance is not always easy. Keith's ribbon controller usually had a hundred feet of cable, but on one occasion he was limited to twenty-five feet and he forgot how far he could go. Will Alexander, Keith's keyboard technician, recalls what happened next: "During the Japan tour, the band was playing in Osaka at a baseball stadium. They were set up in the outfield and the audience was 150 feet away, up where the home plate was. So Keith started playing with the ribbon controller and he went running at the audience, when all of a sudden he reached the end and it knocked him down. It made the Moog go berserk."[7] Keith, always the consummate showman, was able to deal with the incident with aplomb: "He got up, turned and bowed to the audience, and went running back to the stage."[8] By, in effect, turning the ribbon controller into a guitar, Emerson and his audience (mainly made up of young men) were reproducing all the cultural and gender symbolism that the guitar as "technophallus" in rock music evokes.[9]

Music seems to be one of the most conservative areas of cultural produc-

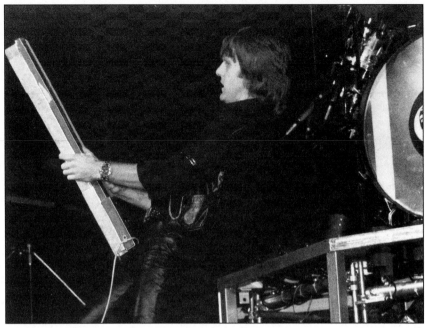
Figure 10. Keith Emerson with ribbon controller, 1980

tion. Here was a new instrument, the synthesizer, one of the few new instruments ever to come along, and people seemed obliged to perceive it in terms of instruments with which they were familiar, the piano and guitar. Escape from these shadows would be difficult.

○

The Moog Ladder Filter

Moog's voltage-controlled oscillators, amplifiers, filters, and envelope generators were initially based on standard circuits. Sometime during late 1965 and early 1966 Moog came up with a novel design for a filter, known as the low-pass filter or ladder filter (after the ladder of transistors in the circuit). This filter is the crown jewel of the Moog synthesizer—it is the "rich," "fat," "juicy" tone that nearly everyone refers to as the Moog sound. This is the filter that Moog's rivals at ARP and EMS were most envious of and which they tried to copy.

Filters, which control the higher harmonics of sound, were not new; they had been used since the days of radio. Bob seems to have designed his unique low-pass filter from a combination of book knowledge and his usual tinkering. He presented the design in a paper delivered to the AES convention on October 11, 1965. A year later, on October 10, 1966, he filed for a patent, which was granted on October 28, 1969. It is the only item on the whole synthesizer that Moog ever patented.[10]

A low-pass filter can be thought of as a gate in a stream. The higher it is raised, the higher the harmonic frequencies that pass through it—or, more correctly, under it, since it's a low-pass filter. If you speak into a long pipe, your voice will be muffled because the pipe is acting as a low-pass filter; talking into a pillow has the same effect. Where the cutoff occurs—that is, how high the gate is raised—depends on where the cutoff control is set. By varying the height of the gate, the timbre or tone color of the sound can be varied. (Timbre is what allows you to distinguish a pitch played by a clarinet from the same pitch played by an oboe.) By voltage-controlling the gate, the musician can sweep the filter through its range, changing the timbre of the sound by emphasizing some harmonics and attenuating others.

Bob found a novel circuit to do this, using pairs of transistors connected by capacitors arranged in a ladder. This makes the filter balanced, because the signals can go up both sides of the ladder at the same time. The signals enter the bottom of the ladder, and those with higher frequencies find it hard to make their way up the ladder because of the electrical properties of the transistors and capacitors.[11] One of the main factors in the quality of the sound from the Moog filter is a characteristic called the cutoff slope. A filter doesn't actually chop the high harmonics off completely but rather attenuates them. The cutoff slope refers to how abruptly the amplitudes of the high frequencies taper off. The Moog filter has a much sharper cutoff slope than almost any other synthesizer filter.[12]

The filter has many other qualities, such as a sharp resonance around the

SHAPING THE SYNTHESIZER

cut-off frequency. When the filter is overdriven—which means that the amplitudes of the signals going into the filter are too large—it produces a rich form of distortion that is characteristic of the fat Moog sound. Jim Scott, a Moog engineer who worked extensively with the ladder filter, adapting it for use on the Minimoog, commented, "This filter defies analysis. There are lots of subtle things going on that almost defy mathematical treatment."[13] The best analog components in sound nearly always have this quality of not being quite understood and nearly always involve some not quite specifiable resonances and distortions that occur at high frequencies beyond the audible range but that produce audible effects.

When the filter is used with an envelope generator in the bass range, the resonant deep sound is particularly appealing and was soon discovered by synthesists.[14] Over the years it has become a staple of pop and rock music, as has the bass sound of the Minimoog (which uses a similar filter). Bob Moog was himself a witness to the power of his bass sound when he was invited to bring his synthesizer to a New York studio session where Simon and Garfunkel were recording their album *Bookends* (1968). Moog set up the bass sound himself for the track "Save the Life of a Child," which opens with this sound: "One sound I remember distinctly was a plucked string, like a bass sound. Then it would slide down—it was something you could not do on an acoustic bass or an electric bass . . . a couple of session musicians came through. One guy was carrying a bass and he stops and he listens, and listens. He turned white as a sheet." The significance of the Moog bass sound was not lost on this session musician. The Moog not only sounded like an acoustic or electric bass, but it also sounded *better*.

Moog liked to repeat this story; he felt that at last he was "getting somewhere." At last the Moog was finding a home among musicians at large, rather than being an instrument merely for the avant-garde. Session musicians were some of the first to see the writing on the wall; their livelihoods were under threat. This threat was something that the powerful musicians' union would eventually take up.

○

But Is It a Synthesizer?

In the early accounts of his work, Moog refers to "electronic music modules" and a "system" of such modules, but he does not use the word "synthesizer." In 1966 he used the term "synthesizer" for the first time in print, and in 1967 he introduced the "Synthesizer Concept."[15] The classical meaning of "to synthesize" is to assemble a whole out of parts. A synthesizer assembled parts of a sound into a complete sound. Moog, like Buchla, thought long and hard about whether to use the term. The name was associated with the RCA synthesizer, but Moog's synthesizer, unlike the RCA, worked in real time and had a keyboard. Moog and Deutsch debated this a lot. Reynold Weidenaar, a composer based at the Moog factory, joined the debate: "I remember when he told me that was his decision, to call it a synthesizer . . . I said, 'You can't do that, you know, the RCA device is the synthesizer, and everybody's going to think of the RCA synthesizer if you use this word, so you're going to have to think of another word.' And he said, 'Well, no, it's a synthesizer and that's what it does and we're just going to have to go with it.' And so he was obstinate, and good thing, too."

By the time other manufacturers like ARP and EMS got going in the late sixties and early seventies, "synthesizer" had become the standard name. Buchla held out longest against the usage, but even he at some point recognized that this was the name that most people were using.

○

What's in the Moog Name

For a short while in the late 1960s and early 1970s, the Moog became the brand name for any synthesizer, in much the same way that the Hoover was synonymous with vacuum cleaners.[16] Branding—making customers aware of your brand, above all of its competitors—can be crucial for the success of a product, as marketers are well aware.

Moog is a Dutch name and rhymes with "rogue," not "fugue." Many people, including musicians, continue to this day to mispronounce the name. Bob even used to have a placard on his desk telling people how to pronounce his name. Several people unfamiliar with the existence of the real Bob Moog have told us that they assumed the name was made up to resemble the sound of the synthesizer itself—MOOOOOOOOOOOG! Cows moo, but synthesizers moog, and as David Van Koevering (the best Moog salesman ever) once said, paraphrasing the sixties lyric by Jonathan King: "Everyone's gone to the Moog [Moon]." The name not only sounds right, it also looks right in distinctive letters adorning a piece of equipment.

And here Moog was extremely lucky. If his name had been Larry Smith or Dusan Bjelic, it is unlikely that his make of synthesizer would have become the brand name.[17] Naming a synthesizer is no trivial matter, as the Japanese, who would later dominate the market, were well aware. Ikutaro Kakehashi, the founder of Roland, the most successful synthesizer company in the world today, told us that he came up with the name by looking through an American telephone directory. Having found the name, he wrote it on piece of paper and attached it to an organ (in those days the company made organs) for a week to see if it felt right. It did![18]

◎

The 900 Series

By October 1965 Moog had standardized the different modules, which became known as the 900 series.[19] The modules varied in price from $195 for a 901 VCO to $475 for a 904 filter. In April 1967 Moog introduced for the first time a catalog with complete "synthesizers."[20] He offered three different models (I, II, and III) ranging in price from $2,800 to $6,200. The most expensive model cost a sizable chunk of money—enough to buy a small house. A more elaborate system with a tape recorder and amplifier could set you back a lot more. Even rock stars were known to balk at the cost. Keith Emerson tried hard to get his for free, and Mick Jagger, on being told the price, famously remarked, "Man, that's a lot of bread."

The first Moog catalog listed numerous customers who had bought Moog equipment, and on the inside back and front covers were endorsements from 21 composers and directors of studios, including Siday, Carlos, Ciamanga, Deutsch, and Nikolais. This impressive list shows that Moog was reaping the benefit of his close links with his customers if for no other purpose than to promote his equipment. The catalog was directed almost exclusively at the electronic music composers who made up the bulk of Moog's customers. "Moog synthesizers are designed to meet the requirements of composers of all types of electronic music."[21] These requirements were "based on discussions with over 100 composers." The first requirement was that "the synthesizer should perform all of the basic generating and modifying operations of the classical studio, and provide additional resources with the state of the art."

Moog had the name, and by 1967 he had the product. Also, the culture around him was slowly changing, becoming ever more receptive to his innovation. We now turn to look at how this culture was experienced in one funky factory, and how Moog set out to change it.

4

The Funky Factory in Trumansburg

To come here was like funk city, you know, you opened the door,
and you stepped in and the floor creaked.

DAVID BORDEN

LOCATED ON THREE floors on Trumansburg's main street,
next door to Kostrub's luncheonette and down the street
from Camel's Bar, the former furniture store looked like any other
Trumansburg business—rundown. The atmosphere of Bob's shop was described to us by several people as "funky." Bob Moog was an unorthodox
businessman and liked to "make do." This was no high-tech operation with
a slick sales force; it was Bob Moog and a few workers sitting in a storefront.
Reynold Weidenaar, who worked there from 1965 until 1968, captures what
it was like: "You had this old building that hadn't been remodeled, it
was not very impressive. You had just a lot of tables. You had simple over-
head lighting and people were working with their soldering pencils, and
that was it—it didn't really look anything high tech at all. It looked like a
lot of small-town women sitting there fussing, like they could be sewing,
until you got close enough to see what they were doing." According to
Weidenaar, solder splatters covered the wooden floors and benches; an ac-
rid smell hung over everything.

As business grew, Moog was able to hire more and more workers. He
started with two in 1963; by 1967 he employed twelve; and in 1969 the company for a short while consisted of forty-two personnel. The front office—a

Figure 11. Assembly at Moog's Trumansburg factory

desk, a secretary, a battered filing cabinet and a postage meter—was first located in the basement and then later on the second floor. On entering the front door you walked straight into the assembly area—several large tables in the middle of the room. Benches ran along both sides, where Moog and other engineers designed, aligned, and tested new modules. Later on, engineering moved upstairs to the second floor. There was a machine shop toward the rear of the building. The famous studio was eventually added at the back of the ground floor. But the studio, like everything else, had a make-do feel. Weidenaar: "It was very undependable because whenever he couldn't make a deadline there would be a hole where there used to be a module, and he would ship it out." It seems that Bob Moog discovered just-in-time production long before it became a favorite Japanese management philosophy.

The makeshift nature of the facility can be gauged from Jim Scott, who

THE FUNKY FACTORY IN TRUMANSBURG

arrived there as an engineer in 1969: "If I needed a niche in the wall to park my phone, I just went down and got a hammer and nail and pounded a hole in the wall and cut up some wood and put a shelf in." The atmosphere among the engineers and musicians who worked there was very informal; everyone helped one another and no one thought of this as a nine-to-five job. Ken Fung worked there for a year, first doing PR work for Moog, and then assembly: "They were pretty chummy and informal and excited about what they were doing and at the same time always a lot of talk and speculation about what Moog was up to, what the next invention would be, would it crash and burn, would it succeed."

Bob would be the first to admit that personnel management was not his strong suit. Early on, his wife, Shirleigh, helped out with the bookkeeping. Later, in 1969, Moog employed first Ray Hemming and then (when he left) John Huzar as his general manager, to try to bring some discipline to the operation. These managers—the suits—who had no experience in the electronic music business were disliked by the engineers and Moog studio musicians, who were used to Bob's casual style of management.

The local people increasingly associated the Moog works with sex, drugs, and rock 'n' roll and never accepted Moog or his business into their community, even though he was providing jobs in a depressed economy. They particularly balked at the musicians who came to town. Nearly everyone had stories about the visit of the legendary black jazz musician Sun Ra. His music was path-breaking and his stage performances used ancient rituals and bizarre space costumes.[1] For Sun Ra, "space *was* the place" long before the sixties psychedelic groups wove space into their music. He traveled with his "Arkestra" in a fleet of aged Cadillacs. Jon Weiss describes what happened when Sun Ra plus entourage came to Trumansburg: "He was an old time blues player and he assembled really excellent jazz players and he did this totally far out space music. He came to Trumansburg where the Moog company was back in 68, and this was a fairly rigid, straight-laced, little sleepy New York state town. And here's this bizarre looking black guy with, you know, robes and all this stuff in the local ice cream parlor!"

Some of the engineers and musicians who worked in the Moog factory were hardly orthodox in appearance themselves. Musicians like David Borden and Chris Swansen had the mandatory sixties long hair and beards. Jim Scott, one of the wilder looking engineers, with his flowing locks and big bushy beard, mentioned that local people referred to himself and Chris Swansen behind their backs as "Abraham Lincoln and George Washington."

Part of the difficulty, especially in the early days, was that electronic music was just not part of everyday culture, and certainly not in Trumansburg. Weidenaar experienced this every time he went for a haircut: "I used to dread getting haircuts because I would have to sit and go through this again and again and again. Every time I had an exchange with someone: 'But, Reynold, what is it?'"

Although Moog's factory is surely one of the most significant developments in the history of Trumansburg, there is no reminder to be found there today of the funky factory that changed the face of popular music. Trumansburg with its five churches is today as it has ever been—sleepy.

○

Guitar Days

Moog's small shop in Trumansburg did not turn into a synthesizer factory overnight. Workers we have interviewed recall that, early on, they made guitar amplifiers rather than synthesizers. R. A. Moog Co., in the course of its brief lifetime, made many, many more guitar amplifiers than synthesizers. Bob's own early attitude toward the synthesizer business was, "And it's not like I stopped the shop and said, okay, from now on all we're doing is this electronic music stuff. This was sort of stuff I did for the hell of it while everybody else tried to make the shop go." And what made the shop go was the guitar. Young guys everywhere were discovering that blasting an electric guitar through an amplifier made a pretty good sound—and certainly a loud enough sound to annoy your parents. The electric guitar symbolized teenage revolt. As Jimmy Page of the formative British group the

Yardbirds, and later of Led Zeppelin, remarked, "The good thing about the guitar . . . was that they *didn't* teach it in school."[2]

After the appearance of the Beatles on the Ed Sullivan show, the guitar business in the States started booming, and Bob Moog, in partnership with Walter Sear, saw a chance to be part of that boom. Sear: "It was the time of the Beatles, and the Japanese government got all the woodworkers together in Japan and they began knocking out electric guitars—cheap, cheap electric guitars. But it was too expensive to also ship the amplifiers because its cubic volume. So a number of wholesalers in New York approached me about making cheap guitar amps. So I got hold of Bob, and he started knocking out these cheap guitar amps."

Bob already had some familiarity with amplifiers. When he started his Trumansburg business, he had designed a kit amplifier. But that had been a high-end amplifier, and Sear needed something cheap and nasty. "Bob didn't understand guitar amplifiers. And he built this molded box and hi-fi amplifier. Well, that's not what you want for guitars. You want as much distortion as possible."

Eventually Bob came up with a better design specifically aimed at the guitar market. The chasses were built in Trumansburg and then shipped down to New York by the truckload (along with the speakers), where Sear and a helper would assemble the completed product. The numerous memos and letters exchanged over a two-year period (1965–1967) between Bob in Trumansburg and Walter Sear in New York are almost exclusively concerned with the ups and downs of the guitar business. And the downs far exceeded the ups. Moog and Sear soon realized that making thousands of cheap guitar amplifiers was a very different business from the theremin or synthesizer business.

They started off with small orders of a hundred or so amplifiers with brand names like Segova and Amper. Soon they upped their production as Sear took in more and more orders. By keeping the production runs high, they hoped to keep the price low. The amplifiers certainly were cheap:

$9.65 for the Amper I and $14.50 for the larger Amper II.[3] They sought every possible economy, shaving off a few cents here and a few cents there so that they could turn a profit.

Because Moog was running his guitar amplifier business on a shoestring, he was always on the lookout for cheap parts. One such source, familiar from his hobbyist days, was war surplus. Sear was not impressed with the quality of some of these components: "And one time he bought a batch of surplus World War II electrolytic capacitors. I called him up and said, 'Bob, that was awfully dangerous.' He said, 'Ah, that's fine, I got 'em on rails with DC, I'm rebuilding them.'" Well, of course, the shipment came, and every one of them hummed like a banshee."

On another occasion Moog saw an opportunity to buy a job lot of cheap Japanese jacks; these, however, failed after being used two or three times and resulted in "truckloads" of returned amps.[4] To add to their problems, they were forced to change the brand name of the Segova amplifiers. Sear, in a memo, alerted Moog to the problem: "Segovia threatened to sue because of the use of the name Segova . . . change [the] trade name at once to Sekova."[5] Moog's sense of humor comes through in his reply: "The bit with Andre Segovia is a gasser. I laughed so hard that people began to wonder what was wrong with me. Sekova is an awful name but we will do it."[6]

As time passed, the problems of keeping the production line going, the continuing lack of money, and Moog's commitment to other projects all took their toll. Moog had increasing trouble paying his suppliers. By November 1965, Moog was receiving parts only if he paid for them COD. And by now he had expanded to "five girls doing assembly and three men doing fabrication."[7]

By July 1966 Sear was reporting alarming news after attending a musical instrument convention in Chicago: "The bottom of the amp market has fallen out."[8] Furthermore he was starting to worry that Moog would not be able to fill the backlog of five thousand orders in time. He admonished his good friend: "Bob, either we are seriously in this thing or not. If not, let's get out of it." By November Moog was getting out, selling off the many amplifier parts he had accumulated to a surplus dealer "at a loss."[9]

Moog, in hindsight, sees the guitar amplifier business as having been a distraction. R. A. Moog Co. of Trumansburg, of course, did not become famous for making the Amper or Sekova guitar amplifier. Sear, and to a lesser extent Moog, can today joke about their "adventures" in the "cheap" and "junk" guitar amplifier business. But back then, when Moog was struggling, it was a crucial source of income, and it was invaluable experience in mass production. The last word that anyone would use to describe the product he was to become famous for—the synthesizer—was "cheap."

◎

Walking on Patch Cords

Moog took whatever projects came along. And one early project gave him a chance to work with John Cage, who was gaining recognition as the leading composer of experimental music. Cage was searching for new ways to combine dance with musical performance. After visiting the Trumansburg shop, he commissioned Moog to make a number of antennas that would make percussive noises as dancers approached them. The aim of the work, Cage wrote, "is to implement an environment in which the active elements interpenetrate . . . so that distinction between dance and music may be somewhat less clear than usual."[10] Moog had a somewhat more down-to-earth view of the event: "It was just a whole, huge amalgam of junk. That was the aesthetic."

The piece, "Variations V," was to premiere on July 23, 1965, with the Merce Cunningham dancers at Philharmonic Hall in Lincoln Center. As the day of the concert approached, Moog ran into more and more technical problems. He did not have enough time to make printed circuit boards for the devices, so the components were hand-wired to solder lugs. Moog takes up the story: "It turns out that that whole batch of lugs had defective plating and all the connections came apart over time. And right to the very

Figure 12. Premier of John Cage's "Variations V," Merce Cunningham Dance Company, 1965: John Cage (left), David Tudor (center) and Gordon Mumma (right)

end we had to have multiple people on hand in New York, the night before the performance and the day of the performance, taking every connection apart, scraping off the plating and resoldering."

The set-up also included ten photoelectric cells aligned with the stage lights, triggered when the dancers broke the beams. The photoelectric cells in turn activated ten tape recorders and ten short-wave radios. All the outputs were wired through a special mixer that Max Mathews (the computer music pioneer from Bell Labs) had made for an earlier Cage performance. The mixer had six outputs that fed six speakers spread around the concert hall. In addition, film and video images were projected on the stage. It was a complicated set-up, and even Moog, whose synthesizers would sometimes look like patch-cord jungles, was impressed: "The stage was covered with patch cord[s]. Here's a sixty foot wide stage that had a thick rug of

77

patch cords on it. Cables up and down. It was [a] surrealistic experience for me to be walking on that much of patch cords."

The premier of "Variations V" was plagued with technical malfunctions. At this stage of his career, Cage was engaged in many projects, and rather than collaborating, he ended up delegating most of the work. His comment on Moog, who probably failed to impress him with his just-in-time production methods, was "non-focused."[11] Cage may not have been pleased with Moog, but being part of a concert in Philharmonic Hall at the center of New York City's art world shows just how far Bob had come. Two years earlier he had been a kit manufacturer and scarcely aware of the New York avant-garde. Now he was working with Cage, Tudor, and Cunningham. His 1965 catalog advertised his work with Cage. If nothing else it gave him credentials with the avant-garde crowd. He was thrilled to meet Leonard Bernstein at the concert and work with the legendary Max Mathews. Moog may have been late, but he had arrived.

◎

Why You Don't Want to Ride the Elevator with Bob Moog

Bob and Shirleigh Moog were all too well aware that even with guitar amplifiers, independent commissions, and the new synthesizer business to supplement their theremin business, they were still struggling. But Bob had one insurance policy against complete financial disaster—his PhD.

One of the reasons Moog was late with the Cage job was that he had not finished his PhD dissertation. With his new business to occupy him, he had not found a lot of time to write it. And the chairperson of his dissertation committee at Cornell, Dr. Henri Sack, was getting antsy. Where was his dissertation on ultrasonic attenuation in sodium chloride? Why was he spending all this time making weird electronics instead of writing? Bob, with Shirleigh's support, decided to make one final push at completing his dissertation: "Shirleigh encouraged me. She said with a PhD you'll never starve. Get the PhD and kits out of your system."

One night in July 1965, Dr. Sack's patience finally deserted him. Moog: "Around July of [1965] we were working overtime, trying to complete the work for Cage. It was not going well. One night at about 9 p.m. there were about a dozen of us in the shop sweating away. The phone rang. I answered it. It was Dr. Sack. He said, 'Moog, whatever is not on my desk by 10 a.m. tomorrow is not going in your thesis. Good Bye.' It was a REALLY long night for me, but that's how I got my thesis done."

His dissertation defense was scheduled in the newly-built cathedral to physics on the Cornell Campus, Clark Hall, and for once in his life Bob was early. As he rode the elevator, he was not thinking about solid state physics, the topic of his dissertation; he was thinking about sound and what the resonant frequency of the elevator was. Every object has a natural frequency at which it will resonate—this means that if a sound source is set off near it and hits that natural frequency, the object will suddenly produce enhanced vibrations (resonance). The resonant frequency of an elevator is rather low, so, to find it, Bob started jumping up and down on the floor. Somewhere between the fourth and fifth floors, he hit the right frequency. The elevator suddenly started bouncing alarmingly in time with his jumps and ground to a halt. Four hours later he was rescued. His defense did eventually take place later that day. He passed with little trouble. Bob Moog was now Dr. Moog. He had his insurance policy, but he never really needed it.

◎

The Summer Electronic Music Seminar

Herb Deutsch's involvement with Bob Moog, the collaboration that started it all, stuttered along. In August 1965 he and Bob organized a summer seminar in Trumansburg.[12] They had found their own collaboration enormously rewarding; the plan was now to bring in more composers who would get to play with the latest equipment, and Moog would get to learn more about their needs. It might even generate a few new orders.

Nearly all the twelve participants at the three-week seminar were academic electronic music composers. The youngest was Weidenaar, who was still a music student at Michigan State University. Everyone was housed locally, and Moog threw open his factory. "This was kind of the introduction to the public. He was now open for business." Moog, Deutsch, and one or two participants gave lectures. The main activity, of course, was experimenting with the new equipment—a few basic modules powered by batteries. It might not have been very sophisticated by today's standards, but it was light years ahead of what these composers had traditionally used. The seminar was one of those "once in a lifetime" inspiring events—a chance to be on the edge before anyone knew there was an edge to be on.

It was also an opportunity for the participants to get to know one another better, reflect on the future of electronic music, and of course have some fun. Weidenaar, who seems to have had as much fun as anyone, describes the ambience: "Well, we had sort of the pro-Cage and the anti-Cage people . . . and a lot of very energetic discussion about that . . . we spent a lot of time in the Camel's Bar down the street, you know, talking about musicians, about the new music, about where it was headed . . . there was high excitement. We worked late into the night . . . and we were making crazy music and figuring things out." The workshop ended with a little concert given to bemused Trumansburg residents.

○

Electronic Music Review

Moog, realizing that a new community of electronic music devotees was emerging, soon developed plans to nurture it. As is often the case, need and opportunity coincided. Weidenaar had arrived at Moog's doorstep, and he did not want to go home. He planned to drop out of college to pursue electronic music, and he had some financial support from his parents and a background in publishing. Together, Moog and Weidenaar hatched a plan for him to edit a new magazine devoted to electronic music.

Electronic Music Review was run by Weidenaar from a basement office in the factory. The first issue appeared in January 1967 under the imprint of the Independent Electronic Music Center (IEMC), with Weidenaar as editor and Moog as technical editor. The magazine had a practical focus. It contained equipment reviews, articles on how to use studio electronic music equipment, essays on different genres of electronic music, reviews of concerts and albums, and listings of events, recent publications, and recordings. An impressive selection of engineers and composers contributed, including an occasional piece by Bob. Wendy Carlos, who in those pre-*Switched-On Bach* days was working at Gotham Recording, New York (and was listed in the magazine as a "recording engineer"), wrote a couple of essays about various studio techniques. Everyone pitched in with record reviews.

Weidenaar had expected that the magazine would be a side-line while he worked in the studio, but he soon found he was devoting more and more time to it. The production of the magazine was time-consuming, with Weidenaar doing all the mailings and typing and hand-justifying the type line-by-line. Like most early Bob Moog ventures, it had that make-do feel.

With the magazine being published from the Moog works and with Bob as technical editor, the question arose as to how independent the Independent Electronic Music Center's publication really was. Weidenaar recalls, "We were trying to make this appear as if it were not a promotion or publicity venture of the Moog Company, which met with mixed success. Some people, of course, felt that this being on the same premises was strictly a promotional venture." The first issue was the only one to run an ad for Moog's synthesizers. But one of the new services introduced to subscribers was rental of the Trumansburg studio: the fee was $100 per week.

○

The Studio

The origins of the studio can be traced to the summer seminar, when Moog had built four rudimentary studios for the composers to use. One of these studios became *the* studio, slowly being revamped over time. The

An open invitation to music educators, composers, and performers...

For the past five years, R. A. Moog, Inc. has specialized in the design and construction of electronic music composition equipment, and in the installation of complete electronic music studios. At present, more than one hundred educational institutions use our equipment in the teaching, composing, and performing of music in the electronic medium. Installations range from basic teaching studios to highly sophisticated experimental facilities. These include computer and other forms of programmed control, as well as manual controllers for real time performance.

As part of its educational program, R. A. Moog maintains a complete model electronic music studio which is open for inspection and extended use by those working in the medium. The Moog studio effectively serves as a laboratory where musicians and engineers can exchange ideas, and new concepts in electronic music instrumentation can be tested.

You are invited to visit our Trumansburg studio. Examine and use all of the Moog instruments, as well as Scully tape recorders, MRS variable speed recorders, and other quality instruments for electronic music composition distributed and serviced by R. A. Moog.

Demonstration studios are also maintained by our representatives in Los Angeles and New York City. West of the Rockies, contact Paul Beaver; in the New York City area, see Walter E. Sear. Both are highly qualified musicians and experienced electronic composers, and will welcome the opportunity to acquaint you with Moog equipment.

R. A. MOOG

R. A. MOOG INC., TRUMANSBURG, NEW YORK 14886 ● AREA CODE 607 387-6191
New York - Metropolitan Area: Walter Sear, 304 Avenue of the Americas, New York, N.Y. West of the Rockies: Paul Beaver, 2925 Hyans Street, Los Angeles, California

DESIGNERS AND MANUFACTURERS OF ELECTRONIC MUSIC INSTRUMENTATION

26   MUSIC EDUCATORS JOURNAL

Figure 13. Advertisement for Moog Trumansburg studio, 1968

plan was to bring more musicians to Trumansburg and provide a kind of test laboratory for the latest equipment. To some extent the plan worked. John Eaton, a well-known composer who helped promote an Italian synthesizer, the Synket, spent a summer in Trumansburg, and other composers, such as Don Erb, visited for shorter periods. After the success of *Switched-On Bach*, commercial musicians also showed up.

Eaton is remembered by everyone not only for some wild impromptu concerts but also for one famous dinner party. Shirleigh Moog was an accomplished cook. She went on to produce a recipe book, the *Moog Musical Eatery* (1978), which was published by Crossing Press (founded by Elaine Gil in Trumansburg, a well-known figure in alternative circles in town and a close friend of the Moogs). Shirleigh's cookbook tried to cash in on the Moog name and included not only recipes but vignettes about famous musicians who visited the Moog factory. She liked to host dinner parties, and often Moog's customers joined the family for dinner. On this occasion it was Jon Eaton's turn to cook. He came from upstate New York and had learned some homespun recipes. As the guests ate the main dish, which resembled beef, he asked them all to guess what it was. No one could. Eaton had caught and cooked a six-foot black snake!

One local composer who used the studio was David Borden. With a Harvard degree in composition and an interest in avant-garde music developed during a year spent in Berlin, he had arrived first at the Ithaca City School District as composer-in-residence (1966–1968). He was then hired by Cornell University as composer-pianist for dance. He soon heard about Moog and the new instrument. Borden: "He was always looking for people to come in and use the stuff . . . To me it was like the inside of a jet airplane, and I had no understanding. He quickly gave me a standard three-hour lecture on how it worked . . . I didn't take anything back from that at all. Oh yeah, 'Thank you.' So, he said, 'Why don't you come in tomorrow and mess around.' So I did, and I couldn't get a sound at all." Borden finally realized that he would have to swallow his pride and ask someone what was wrong. He went upstairs to the office of one of the engineers. "He came down and turned the amp on for me!"

Borden had the reputation of being somewhat of a klutz in the studio. Much to his surprise, he found himself playing an important part in Bob Moog's research to refine his synthesizer design. Borden: "I went every day for about three weeks, and about the third week, I noticed that one of the modules was smoking. I could smell it. I thought, 'Oh shit! Now I've really done it!' but I ran out and got an engineer . . . and then there were about five engineers looking at this thing, plus the guy who tested the equipment. Holy shit! And he said, yeah, I just burned out a filter, or something. It was no longer usable, and they got Moog down . . . well [I said to myself] 'That's it, I'm not coming back,' but it was the opposite. Moog then came out and said, 'Oh, don't worry about it, forget it,' you know, 'we should just give you a key, and you can come in at night.'"

And David did come in again, night after night. He soon got used to that familiar smell of electronics being fried. Moog told him "whatever happens just leave everything the way it was." David told us, "I was idiot-proofing the equipment without knowing it . . . See, they were coming into contact with someone who had absolutely no technical background, which is me." Moog's reputation for reliability was built upon finding all the ways that

83

musicians like Borden could mess up the equipment. Today, we would call this "beta testing." With the flexibility that the patch system gave—inputs could be put into inputs, outputs into outputs—there was a vast number of ways to make a patch and a vast number of ways to mess up a patch. Borden found them all.

Another person to show up in Trumansburg at this time was Jon Weiss. While attending Antioch College he had heard of Moog through the composer Joel Chodabe. Jon's official title soon became "composer in residence." He quickly learned his way around the new equipment. David Borden remembers, "He was a young kid. And he was quite articulate, very smart, and learned the synthesizers and stuff in about a week." He played a crucial role as a translator between musicians and engineers (similar to the role played by Bill Maginnis at the San Francisco Tape Center). It fell to Jon to explain the intricacies of the vast studio system to visiting musicians and to go out on the road. Moog's complete systems still had no operating manual, but he did include the offer of free tuition packages in the Trumansburg studio. The smallest, System I, gave you one free day of tuition; the System II, two days; and the massive System III, three days. This offer (which did not include travel expenses to Trumansburg) was taken up by a number of musicians.

◎

Electronic Music Review Folds

Although the studio was up and running and Moog was learning from the musicians who used it, the *Electronic Music Review* continued to struggle. The second year's issues contained notably more advertising, including ads by record companies for new electronic music albums. But the revenue from advertising and the very limited subscriptions were not enough for *EMR* to stay solvent. Issue 6 in April 1968 carried a direct appeal to readers: "Your Support is Urgently Needed to help keep *EMR* alive." By this point *EMR* had grown from about 500 to 1,500 subscriptions, including many from overseas, showing the degree of success it had achieved as electronic music started to take off. Weidenaar tried to interest a couple of university presses in taking over the magazine, but it was not academic enough for them. With the continuing financial problems, and Weidenaar himself feeling the need to get back to school, he decided to quit and moved to Cleveland. The last piece in the last issue in July 1968 was a rave review by Weiss of Pierre Henry's *Panorama of Experimental Music, Volume 2: Le Voyage*. Henry's record mirrored in musical form the wheel of life described by the *Tibetan Book of the Dead*. But the wheel of life for *EMR* had ground to a halt.

The magazine was a brave venture and, in a way, ahead of its time.[13] *EMR* was launched just before all sorts of musicians took up the synthesizer in the early 1970s. *Keyboard* magazine, a much slicker publication launched in 1975, aimed at this mass market and became a must-read for enthusiasts. It too reviewed equipment and records, offered tips for keyboardists and synthesists, and carried ads from manufacturers as well as readers. *EMR*'s readers had mainly been composers, and although the record ads in the later issues suggested a more popular appeal, this was not a market Weidenaar and Moog had been after. "It wasn't like slick, you know the magazines like *Keyboard* and so forth, and those kinds of things came in soon after . . . it was mostly a magazine for specialists, and not really for amateurs."

○

Reaching the Metropolis

By April 1967, with the magazine started and the factory studio up and running, Moog had laid the groundwork for an imaginative and ambitious conception of what a manufacturer of electronic music instruments should attempt to achieve. He had a product, a catalog, and a price list. Along with his own factory, studio, and journal, Moog's position as a leading innovator in the field of electronic music was assured. He had even been invited to

85

chair the sessions on electronic music at the forthcoming spring 1967 AES meeting. In three years Moog had gone from being a nobody at the AES to becoming the leader of the field.

But there was one last piece in the jigsaw puzzle to be put in place. Although he had the factory and studio, Trumansburg was still Trumansburg. To be really effective he needed to be in the metropolis. He soon found a way of arranging this as well; he suggested to Walter Sear that he set up a studio in New York City: "I hope you have been thinking some more about the idea of your setting up a studio in your office . . . New York City has about as little electronic music equipment per capita as any place I know of. Not only would your studio serve as a provider of rental income, but would also serve as a dandy show-room should you ever be interested in selling our equipment in the metropolitan area."[14] In the same letter (written in early 1967) Moog offered Sear the East Coast dealership, which Sear accepted.

An over-riding difficulty facing Moog's business was a lack of capital and no sense of anything resembling a business plan. Bob Moog's theme song was "Air on a Shoestring." His resources were stretched to the breaking point. The junk guitar amplifier business, though not lacking for customers, had been a financial disaster. The synthesizer business had the opposite problem: a high-end price but too few customers. There were only a few electronic music studios to which he could supply equipment, and by 1967 he had covered most of these. Although he and Deutsch were working on a cheaper synthesizer for the school market—known as the "Ed. Moog"—and he planned to attend an educators' convention in 1967 in St. Louis, the modular systems were prohibitively expensive for most schools. Where were the new customers going to come from?

○

Good, Good Vibrations

The first indications of where this new market might be found came, surprisingly, from the theremin. In 1966 Brian Wilson produced one of the most influential pop singles of all time, "Good Vibrations," and it used a form of theremin. The wailing of the theremin can be heard throughout, and the final cadence of the chunky cellos set against the theremin is a lasting legacy to Brian Wilson's genius. The instrument used on "Good Vibrations" is actually an electro-theremin specially developed by a session musician (a trombonist), Paul Tanner, and Bob Whitsell in 1958 so that its pitch can be changed by sliding a controller along a surface. With the success of this sound, the Beach Boys now needed something to use in live performance, and they visited Walter Sear in search of solutions.

Sear takes up the story: "Anyway, the Beach Boys came around and I showed them the theremin . . . they got up and they tried it and of course all they're getting is whoops and whops out of the thing . . . They said, 'Oh, no, this will never do, we're guitarists, this has got to have something like a guitar fret board.' In those days I don't even think I got on the phone—we used to write because phone calls are expensive. And Bob came back, said, 'Alright, I can build something.'" The actual device Bob built was a form of ribbon controller. Sear remembers showing the Beach Boys how to play the ribbon: "I had a white grease pencil and marked A, A sharp, big lines on this ribbon so they could play the damn thing."

The success of "Good Vibrations" produced a completely unexpected revival in the theremin business, and Moog now geared up to manufacture his different models, at one point noting, "The pop record scene cleaned us out of our stock which we expected to last through Christmas."[15] Moog told Sear about an article on "teeny boppers" he had read in the *New York Times* Sunday magazine where reference is made to "Lothar and the Hand People." This New York group (Lothar was the name of their theremin and the musicians, which sometimes included Bob Margouleff, were the "hand people") was one of the first sixties' psychedelic groups to take up the theremin. They later bought a Moog synthesizer. Moog suggested trying to get a picture of the group and their theremin for publicity—"it might be worth a few hundred instruments—then again it might not." This was the first time, as far as we can tell, that Moog had contemplated using a pop musician in his advertising material.

○

High on the Hills

With the sudden interest in electronic sounds in pop music, several strands were now starting to converge. Paul Beaver, a Hollywood film music sound specialist with a collection of electronic instruments to rent, and his collaborator, Bernie Krause, came to visit Moog to buy a synthesizer. Beaver and Krause planned to use their instrument on a new pop Moog record being made out on the West Coast. Furthermore, Paul Beaver wanted to become Moog's West Coast sales rep. From the letter to Walter Sear (written in early 1967) where Moog offers him the East Coast dealership, it is clear that Moog was pinning his hopes on this new market among pop musicians: "The time could not be more right. From my vantage point high on the hills of Trumansburg, I see a full blown fad in pop electronic music about to erupt. Several musical instrument makers have already come out with electronic sound modifiers. A couple of pop records have already been made with electronic sound and many more are about to come out."[16]

Two pop recordings inspired by this fad were *Kaleidoscopic Vibrations: Spotlight on the Moog* (1967) by Jean-Jacques Perrey and Gershon Kingsley (using Moog and ondioline) and *The Amazing New Electronic Pop Sound of Jean-Jacques Perrey* (1967) by Jean Jacques-Perrey (using Moog, ondioline, and Ondes Martenot). Would the "full blown fad" be anything more than a fad? At the West Coast spring AES meeting of 1967, Moog would find out. Moog remembers that before leaving for the meeting he and Shirleigh decided to give the synthesizer business this "one last try."

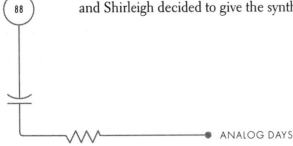

5

Haight-Ashbury's Psychedelic Sound

The music suddenly submerges the room from a million speakers . . .
a soprano tornado of it . . . all-electric, plus the Buchla electronic
music machine screaming like a logical lunatic.

TOM WOLFE

KEN KESEY STARTED it all. He's the one who made the
fateful connection between weird sounds and psychedelic
drugs. It was he who discovered that electronics was every bit as effective
as psychedelic drugs for blowing your mind. The two together were lethal. But Kesey was neither an engineer nor a musician. He was an author, adventurer, and frontier pioneer, especially the frontiers of the mind.
The technical expertise and the connection to the avant-garde world of
electronic music would have to come from elsewhere. It did—from Don
Buchla and Ramon Sender.

San Francisco in 1966 was one of the most exciting places on the planet.
The Vietnam War demonstrations and Civil Rights marches had brought a
generation of young people together, but something new was about to be
added to Berkeley activism and the protest songs of the folk movement.
That new ingredient first crystallized across the bay from Berkeley, in the
Haight-Ashbury district of San Francisco. And the San Francisco Tape
Center on Divisadero Street, at the eastern end of Haight-Ashbury, was
right in the thick of it. Its reputation for staging wild art events and happenings was growing.

Ramon Sender, electronic music composer and one of the Tape Center's founders, was increasingly drawn to the artists, rock 'n' rollers, and dissident young people who made up the Haight-Ashbury community. And members of that community were increasingly drawn to the Tape Center, whose large performance space was the magnet. It attracted the hippest of the hip groups, such as the Charlatans, the house band in the first authentic hippy saloon, the Red Dog Saloon in Virginia City.

Around the corner at 1090 Page Street another big scene was developing. Sender remembers the long stoned night when the name "Big Brother and the Holding Company" was suggested for the group of itinerant musicians living there. "And we would go over there, they would come over. And next door to the Tape Center, Darby Slick's band [Great Society, with vocalist Grace Slick] . . . were rehearsing upstairs . . . In the middle of concerts—it was a cello recital with the Great Society accompanying [from next door]. And there were all these cross connections. Chet Helms was in and out."

It was Helms who discovered that money could be made from the new music. He started charging 50 cents to listen to the famous all-night jams staged by Big Brother in the handsome Victorian building's basement at 1090 Page. Rather than putting people off, he found that a cover charge only attracted more interest. Acid rock, with its free form improvisation, was being born. Charles Perry in his history of Haight-Ashbury comments, "In a few weeks the sessions went from a situation of up to thirty musicians and an audience of eight or ten to one where there were eight or ten musicians and an audience of three hundred."[1] Chet Helms, along with Bill Graham, were to become the new impresarios of acid rock, San Francisco style, with Helms running the Avalon Ballroom and Graham the Fillmore.

The ambience of rock 'n' roll was starting to mingle with the more arty avant-garde and multimedia happenings that the Tape Center specialized in. Theatre was another important ingredient, and the "new theatre" with its audience participation and primitivism, aiming for Zen spontaneity, was a further spur to the happenings at the Tape Music Center.

Light shows were newly emergent at this time. Along with Tony Martin, who worked with Ramon Sender at the Tape Center, there was Bill Ham, an abstract expressionist painter who lived in the Haight on Pine Street and whose shows "were like moving abstract paintings projected in brilliant colors on a screen."[2] The use of devices for synchronizing and modulating light in response to sound, liquid drop techniques, stroboscopic effects, day-glo paint, and black light (UV) all played a part in the new light-show aesthetic. The technology of light was merging with the technology of sound.[3]

And of course there were psychedelic drugs galore. The drug culture and the early San Francisco minimalist scene soon merged. Terry Riley's early pieces were inspired by the use of the naturally occurring psychedelic mescaline. Steve Reich turned Sender on: "[Steve] came over with a paper sack full of these odd, shriveled-up looking things. And he said, 'Where is your blender?' And he put these things in the blender, ground them down to powder, packed them into these large double-o horse capsules, and said, 'You're going to take sixteen of them and I'm going to take sixteen of them.' So we did, and it turned out to be peyote. And I'd never, at that point, even smoked pot . . . And we fooled around with the piano and got high."

Drugs, especially marijuana, were by this point routine in San Francisco. As Bill Maginnis told us, "You didn't have to seek it out. You had to seek out a place where it wasn't." The Tape Center itself tried to discourage the open consumption of drugs on the premises because it had to meet fire codes.

◎

The Acid Tests

The arrival of the most potent psychedelic drug of them all, LSD (lysergic acid diethylamide), into Haight-Ashbury was spurred by two people: Ken Kesey and Augustus Owsley Stanley III, better known as Owsley. Kesey was the famous author of *One Flew Over the Cuckoo's Nest* (1962). He had be-

come an advocate of LSD (coining the name "acid"), and he and his group of associates, the Merry Pranksters, gave out free LSD at events called acid tests. Owsley, the grandson of a Kentucky senator, was a former radar technician who had discovered the pleasure and fortune to be had from manufacturing drugs.

Kesey had discovered LSD as a volunteer experimental subject, back when the drug was legal. His famous trip across America in a bus called "Further," driven by Neal Cassady of the Beats, was immortalized by Tom Wolfe in his *Electric Kool-Aid Acid Test.* Kesey and his friends' approach toward LSD was very different from that of the ex-Harvard professor and East Coast acid guru, Timothy Leary. For Leary, an acid trip was akin to a mystical experience, a way of entering the "Doors of Perception." Leary's way of handling the dramatic effects that the drug brought on was to avoid disruption and unexpected events that might produce the "freak out" experience. Kesey's philosophy of "freak freely" was "to confront fear itself by courting the unexpected."[4]

Kesey had already shown that he had no fear by co-opting the California Hell's Angels to his cause. They gave him the term "bummer" for a bad trip (it used to mean a bad ride on a motorbike). Kesey and friends delighted in "setting up puzzling, unexpected and downright edgy situations to see what would happen. They called these existential practical jokes pranks and . . . started calling themselves the Merry Pranksters."[5] Their most famous pranks involved staged events—the acid tests—where copious supplies of Owsley's LSD was served up in food or drink (hence the electric Kool-Aid acid test).

Sound was an important part of the acid tests. The ingestion of psychedelics can produce strange distortions in the way sound is experienced. Albert Hoffman, the Swiss chemist who discovered LSD in 1943, first noted the synthesia produced by LSD—"every sound generated a vividly changing image, with its own consistent form and color."[6] Kesey's "Psychedelic Symphonette" performed Chinese-sounding aleatory music on flutes and guitars at the acid tests. The pranksters banged on the "thunder machines" of sculptor Ron Boise, whose pieces (like Jason Seley's) were designed to make interesting sounds when struck.

The standard set-up was to have the Warlocks (later renamed the Grateful Dead) playing at one end of the room and Kesey and his musicians performing at the other. Perry describes the first acid test: "Colored light . . . Kesey . . . taking the microphone to make everybody's trip as weird as possible: 'The room is a spaceship and the captain has lost his mind.' Speakers and microphones wired through tape recorders for echo effects or delayed playback . . . The Pranksters playing flutes and guitars in a random way . . . And of course acid, plenty of acid."[7]

○

The Lag

Kesey was also into tape. His bus was equipped with a PA system (including units built by Buchla) and tape recorder; the pranks and antics on the bus ride across America were videotaped as the *Bus Movie* and often played back at acid tests. Kesey used tape and tape manipulation for a more radical purpose than making electronic music or documenting events. He wanted to disrupt ordinary reality, to make people aware that they were living in a kind of existential movie, *the moment*. Wolfe: "By nightfall the Pranksters are in the house and a few joints are circulating, saliva-liva-liva-liva-liva, and the whole thing is getting deeper into the *moment*, as it were, and people are working on tapes, tapes being played back, stopped, rewound played again." Kesey called a tape delay "the lag"; the lag was a way to get to the moment. As Wolfe described it, "Out in the backhouse he has variable lag systems in which a microphone broadcasts over a speaker, and in front of the speaker is a second microphone. This microphone picks up what you have just broadcast, but an instant later. If you wear earphones from the second speaker, you can play off against the sound of what you've just said, as in an echo. Or you can do the things with tapes, running the tape over the sound heads of two machines before it's wound on the takeup reel, or you can use three microphones and three speakers, four tape recorders and four sound heads, and so on and on, until you get a total sense of the lag."[8]

The theme of madness and craziness and questioning the line between the ordinary world and the world of the insane also drove Kesey's vision. If you were prepared to question sanity itself, you could make self-discoveries—expand your consciousness, blow your mind.

○

The Trips Festival

The acid tests were getting bigger and bigger and more and more notorious. This suited Kesey fine. He had one planned for the Fillmore auditorium for early January 1966, but he wanted to get even grander. As he looked for a suitable venue, he ran into an old associate, Stewart Brand.

Brand was an ex-biologist who later went on to found *The Whole Earth Catalogue*; and much later (in 1985) he played a role in forming one of the first worldwide computer links with the 1985 Whole Earth 'Lectronic Link (WELL). Back in 1966 he was staging a traveling multimedia show, "America Needs Indians." Kesey talked to him about his ideas for a much bigger public acid test, and they agreed to join forces. Around the same time, Sender was looking to do something radically different at the Tape Center. Sender remembers getting a call from Brand just after he had taken an acid trip at Esalen, Big Sur. They invited him to help run an event they were planning to call the Trips Festival. It was to feature an acid test, rock groups, a big public celebration of the Tape Center activities, and the Open Theatre (a satirical cabaret).

The festival fell at a bad time for the Tape Center, however. It was negotiating with the Rockefeller Foundation for a new grant, one that would support the organization for several years. It needed a university affiliation, and being associated with an acid test could damage the center's funding

prospects. Subotnick, in particular, was increasingly nervous about the more radical direction in which Sender wanted to take the Tape Center. To keep the pressure off the Center, Sender agreed to formally leave it for a month to run the Trips Festival.

The festival was planned for a weekend, January 22–23, at the Longshoremen's Hall, near Fisherman's Wharf. It was a large, domed amphitheater with appallingly bad acoustics. The festival was promoted as a "non-drug re-creation of a psychedelic experience"; handbills referred to a trip as "an electronic experience." The electronic dimension was further emphasized in the two posters for the event: "a large silk screen one showing a spark, and a cheap black and white handbill featuring an oscilloscope pattern surrounded by an op art swirl." The audience were invited to "bring their own GADGETS (AC outlets will be provided)." Sender saw an opportunity to use the Tape Center's own latest gadget, the Buchla Box: "I invited Don to get involved in the Trips Festival . . . Now what I wanted to do was run Big Brother through Buchla's box, and I wanted to just sit there and crank the ring modulators up, very, very slowly and then just take them out further and further and further."

Soon Buchla found himself doing most of the electronics for the Trips Festival, including building the PA system. Buchla: "So we provided outlets for everybody to bring their toys and plug them in. They could be projectors or musical instruments or anything they wanted, and then we could program everything. It was a gathering of the tribes as we called it . . . the gathering together of musicians to play the spaciest music they possibly could."

The Trips Festival was given a big build-up, starting with a New Year's Eve parade led by Kesey's bus through the heart of San Francisco's financial district. Gadgets were used to advertise the event. Sender recalls adapting the device that Buchla had designed as an aid to blind people and placing it in the City Lights (a well-known book store) window. "Depending on how close or how far people stood at the window, it would play a little melody. It was fun to watch people kind of realize it was reacting to them and their presence. And then it had a big poster for the Trips Festival."

Another high-visibility publicity event took place in Union Square. Kesey had just been busted on Stewart Brand's roof for possession of a big bag of marijuana. The judge warned Kesey not to go to the Trips Festival. Perry: "Kesey and his friends took their garish bus directly from the courtroom to downtown Union Square to publicize the Trips Festival . . . They spoke to the press, set up and played a Thunder Machine."[9] Sender made a big sign saying "NOW" and "we filled a couple of weather balloons with helium and we raised this sign up, we just let it go up in the sky."

With all the attention, the Trips Festival was threatening to get out of control. Sender decided to call up Bill Graham. "Bill was terrific, like he always was. He pulled the thing together, he had the tickets, he had the posters, the door covered, he had his security, everything." Bill Graham, former business executive and member of the San Francisco Mime Troupe (for whom he had just staged a hugely successful benefit) was renowned for his organizational abilities. With his involvement, the Trips Festival garnered even more attention. The plan was that on Saturday night before 10 p.m. there would be various events, including the Ann Halprin dancers and a Vortex Light Box. "Sound would come from the synthesizer invented by Donald Buchla, which would perform on its own and also modulate the rock 'n' roll sounds of Big Brother and the Holding Company in freakish and avant-garde ways. The Acid Test would follow at 10 p.m."[10]

Despite the ban, Kesey showed up at the festival disguised in a spacesuit, carrying his brother's ID. It worked; Space Cadet-in-Chief went around unrecognized by the cops. Dressed in his space suit, he was a domineering presence, as was Bill Graham dressed in a V-neck cardigan sweater. The Dead played when the spirit moved them or else ate the LSD-spiked ice cream or played the Thunder Machines. "THE OUTSIDE IS INSIDE. HOW DOES IT LOOK?" Kesey scribbled on a large screen projected against the wall.

And throughout it all, perched on a big tower in the center of the hall, was Don Buchla running his Buchla Box—making electronic sounds, processing the sounds of the bands, running slide shows and light shows from the Buchla Box, and keeping all the chaotic electronics of the Trips Festival going. No one, including Don himself, can remember exactly what he did—because he did everything. It was, after all, the sixties.

The Trips Festival was a spectacular success in terms of numbers and publicity. It was covered in *Newsweek* magazine and was one of the first manifestations of the San Francisco scene to make it into the wider public consciousness.[11] The event made money, too, lots of it. The day after the festival, Bill Graham signed a lease on the Fillmore. Tom Wolfe saw the Trips Festival as a kind of culmination of the many strands of the countercultural psychedelic experience that were brewing in the Haight:

> The Acid Tests were the *epoch* of the psychedelic style and practically everything that has gone into it . . . it all came straight out of the Acid Tests in a direct line leading to the Trips Festival of January, 1966. That brought the whole thing out in the open. "Mixed Media" entertainment—this came straight out of the Acid Tests' combination of light and movie projections, strobes, tapes, rock 'n' roll, black light. "Acid Rock"—the sound of the Beatles' *Sergeant Pepper* album and the high vibrato electronic sounds of the Jefferson Airplane, the Mothers of Invention, and many other groups—the mothers of it all were the Grateful Dead at the Acid Tests.[12]

The possible effect of this mother of all "electronic experiences" upon Silicon Valley, future birthplace of the electronic revolution in communications, has been noted by Carol Brightman, who points out that Steve Jobs—co-founder and CEO of Apple Computers ten years later—was in attendance.[13]

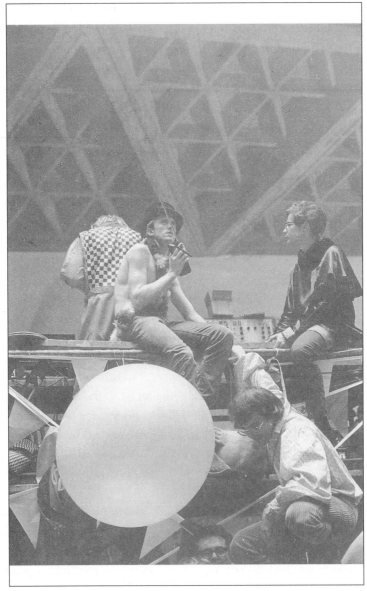

Figure 14. Trips Festival, 1966: Stewart Brand (left), Don Buchla (right)

Figure 15. Buchla's Music Box in use at the Trips Festival

Acid Test Graduation

The last acid test of all, before Kesey served his jail term, was billed as the "Graduation." The Buchla Box was again present (played by the Anonymous Artists of America—a group who inherited all of Kesey's sound equipment). Tom Wolfe, in inimitable style, describes this final acid test where Kesey handed out the graduation certificates: "They're dancing clean out

Figure 16. Ken Kesey's Buchla Box

of their gourds, they leap, they flail their arms up in the air, they throw their heads back, they gyrate and levitate . . . they're in a state . . . they're ecstatic . . . The Pranksters . . . take to the bandstand, all electrified, and they start beaming out the most weird loud Chinese science-fiction music and cranking up the Buchla electronic music machine until it maneuvers itself into the most incalculable sonic corner, the last turn in the soldered circuit maze, and lets out a pure topologically measured scream."[14] The Buchla Box was now a part of this non-university, this technological system for producing outrage and ecstasy. The course in psychedelia had been passed; the class was certified.

○

Whatever It Is

Buchla himself would sometimes turn up at the various gatherings of the tribes to play his own Buchla Box. One such occasion was another Stewart Brand event, a smaller rerun of the Trips Festival held at San Francisco State College on October 1 during the middle of race riots. "The Awareness

Festival" or "Whatever It Is" included America Needs Indians, the Grateful Dead, a Bill Hamm light show, the Thunder Machines, and a team of conga drummers who played in a nearby flea market for fifteen hours non-stop. The finale to the show was staged by Don Buchla on Buchla Box, assisted by Stewart Brand. Perry describes it this way: "Around midnight, Brand staged an atomic apocalypse with Don Buchla . . . They announced to the crowd in the auditorium that Russian missiles, presumably carrying nuclear warheads, had been detected on their way to the West Coast; they had evaded our anti-ballistic missile defenses; they were now two and one half minutes away; two minutes; one minute; fifteen seconds, ten, nine eight, seven, six, five, four, three, two, one—and all at once hundreds of flashbulbs went off as the house lights were cut. All good fun amongst acid-heads."[15]

What an extraordinary event this must have been. The technologies of doom and destruction, the nuclear missiles, were being reclaimed by the forces of peace and love. The ultimate Cold War psycho-horror could be dramatized and tamed on the virtual audio stage that the synthesizer provided. This is the transformative power of the new technology at work. This recreation of the virtual soundscape of war was a taste of what, for a later generation, was to become routine in video arcades and computer games around the world. The sounds and lights of war could now be turned into a virtual experience, materialized, localized, and controlled by one piece of technology.

Explosions, sirens, and rockets were some of the easiest sounds to create on the early synthesizers. But many other effects, like insect noises, bird-song, and space sounds, were also now possible from this little box. And when the audience was "stoned out of its gourds," the experience could be overpowering. For those not used to the machine, this new sonic experience defied description. Buchla enjoyed amazing people with his invention. Sender recalls one such occasion in 1967 when Buchla drove up, with one of his systems, to the Digger commune he was living in (the Diggers were named after a seventeenth-century English sect of religious communists). "We put the speakers up in the top of the orchard and were playing all these Martian sequencing sounds, and hippies would emerge out of the bushes, stoned on acid, convinced that the UFOs had landed, staring in amazement." The sounds of space and space aliens, later such a crucial part of movies like *Star Wars* and *Close Encounters of the Third Kind* and video games like *Space Invaders* had their precursors out there "in the garden." Little did these stoned hippies know, but they were hearing for the first time the birth cries of what later would become a global industry.

○

The End of an Era

Ramon Sender quit the Tape Center soon after the Trips Festival. He dropped out to pursue his mystical visions and to explore alternative lifestyles in various communes. But eventually life in the communes proved too hard and in an ironic twist of fate he turned to Don Buchla for support. By this stage Buchla was running his small shop from his Oakland studio warehouse. "So I was sleeping there, working for him in the daytime and fooling around in the studio at night . . . he had a lot of white lightning acid from the Grateful Dead that we tried out."

The Tape Center did get its grant ($200,000 over four years) and moved to Mills College, along with Bill Maginnis and Pauline Oliveros (as director). Buchla's original System 100 was installed there and is still used to train students in "knobs and wire" synthesis. Mort Subotnick moved to New York and took with him a version of the Buchla 100, which he used to produce his acclaimed electronic music albums. With the more normal bureaucratic constraints of an academic institution, and with the departure of Subotnick and Sender, the Tape Center reverted to being just a normal avant-garde electronic music studio. The trips and happenings were in the past—Ken Kesey did his jail term and moved on, to a farm in Oregon—but the harvest was not over.

In fact, psychedelic sound was just getting up to speed, heading out from San Francisco on the hot winds of change that blew through the sixties. The bands that emerged from the Trips Festival kept on truckin', and there was no bigger and no more psychedelic a band to hit the road than the Grateful Dead.

◯

The Grateful Dead

The Dead were one of the first of the San Francisco groups to realize that sound itself was the key to the psychedelic experience. Eventually the Dead acquired what was widely recognized to be the best sound system in the business, their "wall of sound." They experimented with the way sound is made and with electronics.

It was Owsley who had first "seen" the Dead's sound while on an acid trip and had decided that they were "going to be greater than the Beatles." As Brightman comments, "Owsley no longer saw himself as the electrical engineer he had trained to be, nor as a chemist, but as an *artist*."[16] Unlike the Dead, Owsley had the money (profits from his LSD operation) to nurture his artistic fantasies. "There was but a hop and a skip between Owsley's fevered imagination, wired for sound and hungry for a recognition that his secret trade denied him, and the Grateful Dead's inexhaustible appetite for drugs, amps, and synthesizers."

After the Trips Festival, Owsley paid for the Dead to stay in a house in Los Angeles, his latest acid factory. There he started to rebuild their sound system. Brightman, whose sister was the Dead's lighting engineer, writes: "The mom-and-pop production figures for the band in the mid '60's, reported in the *Grateful Dead Family Album*, skirt the hidden costs of the insatiable appetite for new gadgets that Owsley (or 'Bear' as he was called) stimulated in the band—synthesizers that could make music out of any sound and a stereo PA system years ahead of its time." Owsley (Gyro Gearloose was another nickname) commissioned Don Buchla to help

103

build the Dead's PA system and make several synthesizer modules (all colored red).[17]

The Dead were particularly receptive to the use of electronics, not only because of their own experiences with the acid tests and Trips Festival but also because Phil Lesh, their bass player, had earlier studied electronic music with Luciano Berio at Mills College. Through Lesh the Dead eventually acquired their first regular synthesizer player, Tom Constanten, who was taking the same class from Berio (as was Steve Reich). Lesh and Constanten shared an apartment in 1962, described by Constanten as "an avant-garde music factory."[18]

Constanten (along with Terry Riley) studied for a summer with Stockhausen and after service in Vietnam hooked back up with Lesh in 1967, who by this time was playing bass guitar with the Grateful Dead. The band was about to make its second studio album, *Anthem of the Sun* (1968), and Constanten was invited to participate, adding some prepared piano and other studio electronic effects. Constanten stayed with the Dead for two years, including Woodstock, and played Moog synthesizer on the their third album, *Aoxomoxoa* (1969).[19]

Buchla himself has fond memories of this period, including his early relationship with members of the band. (Buchla also had his own band during this time, Fried Suck.) As well as helping build part of their sound system, he would sometimes play along with them on his Buchla Box, occasionally from back of the stage, and more often from the sound/light booth, where he frequently directed the sound mix. I didn't know at the time that one wasn't supposed to sneak up on stage and play an instrument. It looked to me like all they were doing was tuning up anyway."

◎

The Acid-Rock, Progressive-Rock, Art-Rock, Space-Rock Underground

The exploration of sound, the searching for new and unusual sound washes, was to become the hallmark of much progressive, art, space, and acid rock (or "underground" music) in the late sixties and early seventies. The movement was international, with much cross fertilization and manifestations in many different countries and cities. London had a scene similar to Haight-Ashbury, where experiments with tapes, feedback, phasing, pedals, and filtering, plus the addition of unusual instruments and orchestral effects, played a part in nearly all the British bands who were to become household names.[20] Many instruments and techniques besides the synthesizer were involved, but whether it was a theremin added to Captain Beefheart and the Magic Band in tracks like "Electricity" from the album *Safe as Milk* (1967), or Jack Bruce playing through a fuzz box on Cream's album *Disraeli Gears* (1967), or Pink Floyd experimenting with effects pedals and feedback from Syd Barrett's Fender Telecaster and Rick Wright's Farfisa organ on their album *Piper at the Gates of Dawn* (1967), such explorations had the common thread that it was sound itself which made the difference.[21] And the effectiveness of these sound manipulations stretched well beyond that of the underground genres. For example an R&B influenced pop group like Small Faces could add phasing as they did on "Itychycoo Park" (1967) and have themselves a hit.

The effect of the psychedelic movement occurred in all sorts of unexpected places and ways. Bob Moog (who, like everyone else at the time, was known to smoke an occasional joint) told us that while everyone knew that the big rock stars were tripping out, no one thought of a guy like Eric Siday, the classically trained commercial composer of sound signatures who was in his sixties, being turned on. Bob: "When Siday first knew me well enough to talk about things other than patch cords, he talked about a little trip he took . . . to take part in some religious ceremony of an Indian group in the southwest somewhere, and everybody ate these funny mushrooms, had hallucinations. My point there is that he was aware of and participating in psychedelic stuff, and having participated, he then produced these sounds . . . that are played millions and millions and millions of times and it becomes part of everybody's sonic wallpaper."

We would not want to suggest a cultural uniformity in how these explorations into sound and music were manifest.[22] In many new places and at different times people rediscover what Ramon Sender, Don Buchla, Ken Kesey, and many others first discovered at the acid dawn of the synthesizer revolution. It is exactly the play between local and global cultures that makes each uptake so interesting in its own right. Sound technology—so important as a carrier of global culture—gets reworked and appropriated in new local contexts, sometimes generating new cultural forms that in turn push global technology forward. For this reason, an analysis of progressive rock or psychedelic rock or any other kind of music would be incomplete without a focus also on the technologies used.[23]

An irony in the history of the synthesizer is that it was Buchla's invention, not Moog's, that made the scene at the Trips Festival. But the link between psychedelia and sound in this first appearance was too wild, too radical, too "far out." It was the tamer, more usable synthesizer designs of Bob Moog that were eventually to capture the imaginations of the sixties generation and bring the synthesizer back down to earth. In order for that to happen, the synthesizer had to align itself with yet another technologically mediated domain of popular culture—the recording industry.

6

An Odd Couple in the Summer of Love

Because something is happening here
But you don't know what it is.
Do you, Mr. Jones?
  BOB DYLAN

B OB MOOG ARRIVED in Los Angeles for his first West Coast Audio Engineering Society convention in April 1967. Expecting to greet him as he got off the plane was his prospective sales rep, Paul Beaver, who waited and waited. In those days, first- and coach-class passengers had separate exits, and Beaver was at the wrong exit. Moog, as usual, traveled coach class. Bob likes to tell this story as a way of introducing the difference between the affluent West Coast and upstate New York.

At the convention, held in the Hollywood Roosevelt Hotel, Bob set up the synthesizer he had brought with him. Soon Paul Beaver was bringing his friends in. Moog: "They'd put the phones on and I'd start in, 'Turn this, put this in here.' They'd say, 'Oh, wow! Oh, man! OH MAN! OH WOW!' you know. And then you leave them alone for ten or fifteen minutes, or maybe half an hour, and then the next guy would start . . . they told their friends and on the second day we had a line, four people were waiting at a time to hear this thing. It was unbelievable!" This was the first time these West Coast rock musicians had heard the Moog synthesizer.

Soon Paul (with Bob in attendance) had set up the Moog at a recording studio. Moog: "All these people were going around with their shirts like this [unbuttons his shirt]. With a big, mother-fucking amulet right here [points] and their pot bellies sticking out. And they're talking to each other about how great it is to be on dope. You know, they were all—this kind of LSD over that kind, and the experiences they had—it was, hmm, just bullshit. And they got through a couple of takes and they listened to it and it sounded pretty good—these big speakers and what not. 'Oh, Man!' they said, 'this is real head music. Every head in the country is gonna have to have this.'"

He was at Western Studios, and the record they were making was *The Zodiac Cosmic Sounds* (1967), the first record made on the West Coast to use a Moog synthesizer. This was Bob's first real exposure to the heads and their music. For Bob, the sixties meant something very different: "I was working my arse off, and Trumansburg is a pretty isolated place so that wasn't part of my everyday experience." Unlike Buchla, who lived on the edge, Bob watched the edge: "I wasn't a participant in the culture . . . I didn't have long hair, I didn't have beads, I didn't walk around with platform shoes or bell bottoms. I didn't do dope. And, on the other hand, I didn't pass judgment on it . . . I just watched them and it was sort of fun."

Bob was ambiguous about the heads' newfound enthusiasm for his invention. It was all a long way away from what Herb Deutsch and he had set out to do back in 1964. That scene in the LA recording studio is something Bob likes to revisit. Once, at home, relaxing over a glass of wine, he described it to us in a slightly different way. The unbuttoned shirts, the pot bellies, and the amulets were there, along with the dope and LSD, but there was a new ending to the familiar story: "I thought that was such a crock of shit, you know. I didn't know if I wanted to be, have my synthesizer associated with that."

It may have been a crock of shit but it was also a pot of gold, and *that* was Bob's dilemma. He had once entertained hopes that his synthesizer would find its home in classical music. The person who has probably come closest

108

Figure 17. Paul Beaver (left) and Bernie Krause (right)

to realizing this dream for the synthesizer is Wendy Carlos. But Bob was also a realist who had built his company around the ethos of listening to and responding to customers. If there were new customers, no matter how bizarre they might be, they were still customers. Furthermore, he now had an associate—Paul Beaver—who understood this breed of musician and was eager to be his sales rep. He could let Paul deal with the hippies. And that's what he did. He gave Paul the West Coast as his sales territory and headed back to the less heady ambience of Trumansburg. It was one of the smartest decisions Moog ever made.

Paul Beaver was a talented musician with a deep technical understanding of electronic instruments. He was also part of a team—Beaver and Krause. They played sessions, made commercials, worked on movies, and created some of the definitive Moog synthesizer albums of the period.

AN ODD COUPLE IN THE SUMMER OF LOVE

Today, Bernie Krause lives on a ranch well to the north of San Francisco. He has his own studio and works in bioacoustics, specializing in recording natural sounds. He is still friends with many of the producers, such as George Martin, who shaped the sounds of the sixties. Unfortunately, in January 1975 the person he was closest to, Paul Beaver, collapsed at the end of a concert at UCLA from a massive cerebral aneurysm and died two days later. In the peace and calm of the California wine country, Bernie reflected on his friend, his own career as a synthesist, and the turbulent times of the sixties. He had just published his own autobiography, *Into a Wild Sanctuary*.[1]

○

Bernie Krause

Bernie Krause was a child prodigy who learned violin and studied music from the age of three and a half. He moved on to the guitar, first learning classical and, later, jazz guitar. Then along came rock 'n' roll and that changed everything. He wanted to study guitar at music school but soon discovered that it wasn't considered a proper instrument. He ended up studying Latin American history at the University of Michigan and worked his way through school playing sessions for Motown in nearby Detroit. He became deeply involved as a political activist in the Civil Rights movement.

Bernie eventually went on to play guitar and sing in one of the best-known folk groups of the day, the Weavers. Two members of the Weavers had found him in a Cambridge, Massachusetts, coffeehouse performing satirical versions of their songs. They laughed so much that they asked him to join. Within a few months he was introducing "Guantanamera" to a packed audience in Carnegie Hall as they played a reunion concert with Pete Seeger.

In 1964, after the Weavers disbanded, Bernie moved to Los Angeles. In 1963 he had seen Paul Beaver at a Weavers concert in Santa Monica and had played guitar on some LA studio sessions where Paul was playing "dif-

Figure 18. Mort Garson, 1969

ferent oscillators and different things on a film soundstage." By 1966, the time he actually got to work with Paul, Bernie had moved to San Francisco. He had also moved from folk to electronic music. Dylan's going electric at Newport in 1965 was the spur. He began to explore all sorts of new musical possibilities. Stockhausen, who was giving a series of lectures at the Tape Music Center, was an inspiration, and he decided to enroll. It was there that he had his first experiences with the Buchla Box as a student of Pauline Oliveros. He was captivated. He read an article in *Time* magazine about Eric Siday's success as a composer of "sound signatures." "That's for me," he thought. He jumped on a plane to New York and met Siday, who showed him his Moog synthesizer.

AN ODD COUPLE IN THE SUMMER OF LOVE

In late 1966 he got a phone call from Jac Holzman, the hip young president of Elektra records.[2] Based in Greenwich Village, Elektra was moving from folk to the new electric sound and was signing up some of the most exciting sixties rock acts around, such as the Paul Butterfield Blues Band, Love, and his most inspired choice of all, the Doors. Holzman (who had built radios as a boy) was one of the first in the record industry to see the commercial potential of electronic music. Caught up in the exploration of psychedelic drugs, Holzman wanted to do a record that would appeal to the emerging "underground" culture. At the dawning of the Age of Aquarius and with astrology all the rage, he planned a record with a different track for each sign of the zodiac. *The Zodiac Cosmic Sounds* (with music arranged by Mort Garson) would use electronic effects along with traditional instruments (and poetry).

Since electronic music was Bernie Krause's new interest (and he was a consultant for Elektra) and Paul Beaver was an electronic effects specialist, they were an obvious team for the album. Elektra flew Paul up to San Francisco to meet Bernie.

○

Paul Beaver

Born in Ohio in 1926, Paul Beaver studied classical music and became an organist. He acquired his knowledge of technology while serving in the navy during the Second World War. After the war, at age 19, he took his first gig playing church organ for Aimee Semple McPherson, one of the first radio evangelists who ran a church called Angela's Temple in Los Angeles.[3] Aimee's zany lifestyle suited Paul; she was reputed to ride up to her church in black leathers on a motorbike and scandalized her congregation by having an affair with her radio recording engineer.

Paul had a deeply weird streak in him and loved anyone who was a bit different. He was a scientologist and believed he was a cosmic renegade from another planet who had been dumped on Earth 22,000 years ago—for him Earth was like Australia, the place where all the convicts and renegades were sent. He once asked Bernie to join him on the lower slopes of Mount Shasta to inspect the refueling stations of interstellar vehicles. He sometimes walked naked around the mountains looking for the spaceship entrances. He mostly hung out with people at the margins of society. He ate alone in Denny's restaurant but on occasion he liked to live it up at the Magic Castle restaurant in LA, where magicians would come to his table and perform tricks.

Paul loved organs and was for a while a technical consultant to the Hammond organ company (he owned five specially modified Nova-chords). There was nothing he loved better than to jam with the famous black violinist Papa John Creech, of Jefferson Airplane/Starship, on the ferry between Long Beach and Santa Catalina Island. In the basement of his battered LA studio—a one-story, red brick warehouse—he kept an old Wurlitzer pipe organ. His lifelong ambition was to open a restaurant with a fountain in the middle surrounded by a vast pipe organ. He dreamed he would rise from the center on a platform playing the organ in the midst of happy diners.

By the time Bernie met Paul, he was a successful Hollywood session musician who also rented out instruments from the remarkable assemblage he had collected over the years. The cosmic renegade made music and special effects for many early sci-fi movies, including the electronic score for *The Magnetic Monster* (1953). Moog: "And the first time I met him, which was when we went out there, he had this couple of rooms for living space, but then outside of that was this old ratty warehouse full of all this stuff. And every time that something went out for half a day it was fifty bucks, which back then was, you know, you do that a couple times a day, that was very nice."

○

The Odd Couple

Beaver and Krause had a complex friendship. As Bernie told us, "He was 13 years older than me. But as soon as we met we immediately became friends and colleagues." In many ways they were opposites. Bernie's father, like Bernie himself, was a left-wing radical. Paul's family was conservative, and Paul was a Republican to the right of Nixon; he always wore a wide-lapel, double-breasted blue serge suit with a small elephant adorning his left lapel. Bernie was married throughout their partnership. Paul was single and bisexual. An early proponent of sexual liberation, he sometimes liked to relax after lengthy recording sessions by taking friends to a sex club, inviting them to share with whomever his partner at the time happened to be. Beaver and Krause, although the best of friends, were a very unusual pair.

With the offer from Holzman, they saw an opportunity. If they bought a synthesizer, they could perhaps get enough studio work (commercials, records, and films) to make a go of it. But first they had concerns about whether the newly invented synthesizer would be robust or accurate enough for their purposes. After checking out the Buchlas at Mills College together, they decided to visit Bob Moog.

Krause: "We told him our concerns about the synthesizer, and particularly his synthesizer, that the oscillators weren't stable and the equipment wasn't stable . . . Moog said, 'I'll show you how stable it is' and he set an instrument up on the table and shoved it off—a three-foot table—and it fell to the ground; he set it back up again and he said, 'This is how stable the instrument is.' And, sure enough, it worked well enough."

This demonstration was, to say the least, convincing. It so impressed them that, then and there, they decided to pool their life savings and buy Moog's most expensive synthesizer, the Series III, which was shipped out to LA.

○

The Zodiac Cosmic Sounds

The Zodiac Cosmic Sounds with its combination of orchestration, poetry, and weird sound effects, did quite nicely for Elektra. Not every head

Figure 19. The Zodiac Cosmic Sounds

bought it, but enough did for it to become a cult underground record. To-day, it is a wonderful period piece with its stipulation in red letters on the cover, "MUST BE PLAYED IN THE DARK" and its listing of all the signs of the zodiac of those who made the record. The first track opens with an oscillator slowly increasing in pitch before the studio orchestra comes in. After a sudden pause a voice in appropriately saturnine tones announces: "Nine times the color red explodes like heated blood—the battle is on!"

One person in the UK to be influenced by *The Zodiac Cosmic Sounds* was Decca record producer Tony Clarke. He had been given an early guitar and vocal demo of a song "Nights in White Satin" by the Birmingham R&B-influenced pop band, The Moody Blues. This song and the others on the album he helped produce, *Days of Future Passed* (1967), were given the stamp of rock combined with orchestration and poetry. Clarke was influenced by the Beach Boys' *Pet Sounds* (1966) and the Beatles' *Sergeant Pepper* (1967), but "there was another, more curious album which turned all of our heads, titled *Cosmic Sounds*. It was filled with the most unusual music I had ever heard in my life, full of different moods and very cleverly done, sounding not unlike today's computer music, only the 1967 equivalent."[4] The new Moody Blues sound turned them into international rock stars. The heads and their music were slowly gaining in influence.

○

Looking for Work

Paul and Bernie's newly acquired Moog sat on a long table in Paul's studio. Sometimes, exhausted after a 30-hour session in which they struggled to understand the machine, Bernie would sleep under that table, no doubt dreaming of patches. Paul's familiarity with a range of electronic instruments helped him. He patiently explained to Bernie everything he discovered, and Bernie wrote up notes from their conversations.

They produced some demos such as a Moog version of "Monday, Monday," which the Mamas and the Papas had made popular, and started doing the rounds of the record companies. But no one was interested. Although the heads understood the power of synthesized sounds, the record companies could not yet see their commercial potential. Beaver and Krause were experiencing the same difficulties as Wendy Carlos on the East Coast, who first used her Moog to make a version of "What's New Pussycat." No one yet knew *what* to use the instrument for. "Finally, with the last bit of money that we had at that given moment . . . We rented a booth at the Monterey Pop Festival in June of 1967." Monterey changed everything.

On the flight down to the festival, Bernie sat next to Jac Holzman and showed him the notes he had made on how to play the synthesizer. By the time the plane arrived, Beaver and Krause had a record deal to produce *The Nonesuch Guide to Electronic Music*. Nonesuch was a new record label started by Holzman to market classical music, particularly unusual and baroque music, for a new young audience at a reasonable price (the same price as a trade paperback book) and with slick modern packaging. The project had been called "Nonesuch" because when Holzman had come up with the idea, he didn't want other labels to find out what he was planning, so everyone had been instructed to say that there was "nonesuch project."[5] Holzman was looking for experimental music to give the label avant-garde credibility.[6]

○

Monterey Pop

The Monterey International Pop Music Festival turned out to be one of the great rock festivals of the sixties. The Swinging London set in their latest Carnaby Street gear mixed with cool Greenwich Village types and the Haight-Ashbury flower children for three days of love, peace, music, and drugs. The drug of choice was Owsley's latest batch of LSD. Monterey was a showcase for the San Francisco sound, with groups like Jefferson Airplane, Country Joe and the Fish, and Big Brother and the Holding Company (now with Janis Joplin) starring. Ravi Shankar played an inspired set, and Monterey also launched the Jimi Hendrix Experience (introduced by Brian Jones of the Stones) into America with a mesmerizing display of feedback and guitar burning. Many of the San Francisco groups, on seeing Hendrix and The Who, realized for the first time what was possible if they got a recording contract. Monterey was the place where subculture became mainstream.

Although Lou Adler, the LA promoter of the festival (John Phillips of the Mamas and the Papas was another) had promised a nonprofit event, with posters publicizing it as "Music, Love and Flowers," the reality was somewhat different. D. A. Pennebaker's documentary, *Monterey Pop*, captured the event—at least those artists who agreed to sign the last-minute release for the movie. (The Grateful Dead and Quicksilver Messenger Service refused and thus do not appear in the movie.) What Monterey meant for the future of electronic music was that it brought together three key things: psychedelic drugs, the Moog, and money. And the reason there was money was because the record industry could at last see the effect of the screaming feedback from Hendrix's guitar and the popularity of the new psychedelic sound.

Beaver and Krause set up their synthesizer in an open-air booth on the Monterey fairground. Initially, the headphones attached to the big machine in the tiny booth made no impression at all. Krause: "Little by little, as people got more stoned, they came skulking into the booth. And we'd set things up like thunderclaps and Hammond organs and people were really impressed . . . And they came in all guises . . . Like representatives of the Byrds, a couple of folks from the Beatles were there, and the Stones and others, lots of different groups. And ultimately, many of them, because of the large advances that the record companies were offering, ended up buying synthesizers from us, right there, on the spot. And I think we probably sold six or seven synthesizers at $15,000 a crack at that concert alone in maybe one afternoon. I mean, it was, like, unbelievable."

The link between psychedelic music and the synthesizer, born with the Buchla Box at the Trips Festival, could now be taken up by anyone who had $15,000 to spare. The rock bands, high on drugs and armed with their new record advances, were a perfect bunch of new customers for the Moog.

As word spread, the Moog synthesizer became, almost overnight, the latest toy that an aspiring rock star *had* to have. After Monterey, Beaver and Krause were inundated with customers. They worked for a litany of groups including the Doors, the Byrds, the Monkees, Crosby, Stills, Nash, and Young, Frank Zappa, Van Morrison, and the Beach Boys.

○

The Byrds

A typical project was their collaboration with the Byrds on *The Notorious Byrd Brothers* (1968), the recording of which started a month after Monterey. Based in LA, the Byrds, inspired by watching the Beatles' *A Hard Day's Night* (1964), were determined to become pop stars themselves. They had turned Bob Dylan's "Mr. Tambourine Man" into a huge hit by adding an introduction drawn from Bach, putting it into the Beatles' trademark four-four time, and using a Rickenbacker electrified twelve-string guitar.[7] They were a huge influence in Britain. But the Beatles now had *Sergeant Pepper* out, and the Byrds needed a response.

The liner notes to *The Notorious Byrd Brothers* describes it as "answering recent work by the Beatles and the Stones—tackled subjects as diverse as the agony of the Vietnam War and the explorations of deep space." The spacey ambience of the whole record is generated by the unmistakable sound of the Moog. Throughout, bits and pieces of Moog can be heard, in synthesized dolphin sounds, delicate oscillator trembles, and the strange sound washes with filtering and added echo that achieve the classic spacey feeling.[8] This album came to be seen as one of the most important to emerge from the psychedelic era and is widely regarded as the Byrd's finest album. Prendergast in his survey of ambient music writes that "the Byrds were the first electronic pop group who didn't just use technology to sound better but made electric sounds the very nature of their exploration."[9]

○

The Doors

The LA band of the moment, however, was the Doors. They were psychedelia personified and at the height of their success. "Light My Fire" (1967) had become a massive hit for them. America at last had its answer to the British Invasion. The Doors in-your-face attitude had, however, managed to alienate the promoters of Monterey, and consequently the Doors were not on the bill. But the Doors didn't need to be at Monterey to hear about the Moog synthesizer. After all, they were on Jac Holzman's label.

Beaver and the Krause got the call to Sunset Sound Studios in fall 1967 as the Doors set about recording their second album, *Strange Days* (1967). Bruce Botnick, the Doors' sound engineer, recalls the Doors hearing *Sergeant Pepper* three months before it came out and "absolutely flipping out."[10] The Doors resolved, "Let's not do it the same way we did before, let's invent new techniques of recording. No holds barred." Using a newly available eight-track recorder, their experiments included *musique concrète*, tape effects, and the Moog.

One of the most detailed accounts of Paul Beaver's studio work comes from the Door's keyboard player, Ray Manzarek:

> Paul Beaver . . . began plugging a bewildering array of patch cords into the equally bewildering panels of each module. He'd hit the keyboard and outer space, bizarre, Karlheinz Stockhausen-like sounds would emerge . . . Who knew what he was doing? And then he turned to us, all huddled in the control room, and said, "If you hear anything you want to use, just stop me." "Well, yes," Paul [Rothchild the producer] said. "Actually that sound you had about three sounds back was very usable. Could you go back to that?" "Which sound was that?" said Paul Beaver.[11]

The problem Beaver and other early synthesizer players faced was that none of the sounds yet had names. One way to recognize the sound was to describe it:

> "That crystalline sound," Jim [Morrison] jumped in. "I liked the sound of broken glass falling from the void into creation." "Which sound was that?" said Paul Beaver. "A couple back from where you are now," Rothchild said. "It reminded me of the Kabbalah," said

Jim. "*Kether*, the I AM, creating duality out of the one. All crystalline . . . and pure. You know, *that* sound." "Did I make a sound like that?" "Sure," Jim said. "A couple back." "Just go back to where you were," said Rothchild.

And Paul Beaver began to unplug and replug patch cords, and twist little knobs, and strike the keyboard, which emitted strange and arcane and utterly unearthly tones that sounded nothing like the Kabbalah or *Kether*; the crown of the *Sefiroth*. None of the sounds he was creating sounded pure and crystalline. And then we realized . . . he couldn't *get* back.

This was another difficulty in operating the early synthesizers. Even if you could recognize the sound, it was not humanly possible to remember exactly how you had set up all the patch wires and adjusted the numerous knobs.[12] There was no guarantee you could find exactly the same sound again. That was the beauty and the frustration of analog synthesis:

Finally, with Paul Beaver ripping and tearing at his cords and twisting knobs at an increasingly furious pace, sweat dripping from his forehead, ungodly shrieks emanating from his keyboard, Rothchild shouted out, "Stop! Wait a second. Just stop there." . . . The possibilities were endless. The permutations were infinite. And the Beaver seemed as if he was going to try them all, as we watched, going slowly insane.

"Just stop, Paul. That's a good sound there. I think we can use that." A great sigh of relief emitted from the Doors' group mind.

Once the Doors had breathed their collective sigh and turned back from the brink of madness, they did manage to use the Moog on one or two tracks on the album. On the title track "Strange Days" Jim Morrison's vocal is created by the filter and envelope, triggered by Jim himself hitting the keyboard on the vocal—looking, as Manzarek describes it, every bit "the

121

mad space captain." Manzarek summarized his reaction to the Moog as follows: "What an experience of electronic mayhem. Into the infinite!" The Doors may have prided themselves on opening the psychedelic doors of perception but when it came to the infinite, Paul Beaver and his Moog won hands down.

○

The Stamp of the Synthesizer

The vast possibilities the Moog offered in the hands of a skillful synthesist like Paul Beaver contrasted with the way most rock groups and producers wanted to use it. The Moog was largely seen as a way to add an unusual psychedelic effect here and there, as indeed the Doors had done. Krause: "'Well, we want that kind of phase-like sound that you get with filters, that distance sound, we want a weird sound.' They had no knowledge of what it was we were actually doing. And we'd go through several stages, and depending on how stoned they were they would come up with a decision at some point during the evening . . . I almost never did a session where guys weren't pretty much out of it."

Krause was becoming increasingly disillusioned with the rock groups' antics and the difficulty in communicating the possibilities of the synthesizer. But the work kept pouring in: "Synthesizers became so popular at the time that any music without a stamp of the synthesizer wasn't considered terribly valid." The demand was so great that Beaver and Krause bought a second Moog so they could now go to sessions separately.

Film work had been Paul's specialty. With the capabilities of the Moog finally recognized, Beaver and Krause were regulars on Hollywood soundstages. Their first movie to use the Moog was *The Graduate* (1967). Film work paid handsomely—up to four hundred dollars for a three-hour session. The Moog was ideal for adding special effects. Krause: "By the time I finished work with the synthesizer in . . . 1982 or '83, which was my last film score, *Invasion of the Body Snatchers*, we'd done 135 feature films." The most famous of these was *Apocalypse Now* (1979). There was hardly a synthesizer player on the West Coast we talked with who hadn't worked for or at least auditioned for a part in that synthesized score. As Don Buchla told us: "We all worked on that one."[13]

As the session work for films and records rolled in, Beaver and Krause continued to sell synthesizers. They also started to offer "classes, where people would show up, sometimes 30 or 40 composers at the same time." Suzanne Ciani was one such pupil; she took a class Krause gave in a studio in San Francisco. They also offered individual lessons at night for people who had purchased a synthesizer: their pupils included George Martin, the jazz fusion player Dave Grusin, the Beach Boys, and Frank Zappa. And it wasn't just rock musicians who were interested: one unusual customer was Bing Crosby.

Bob Moog, back in Trumansburg, was all too aware of what was happening. At last he was selling synthesizers in significant numbers—between 20 and 30 modular systems in LA alone. The small trickle had turned into a gushing river, and he was having a hard job fulfilling the orders. Moog: "To cut a long story short he [Paul Beaver] sold a quarter of a million dollars worth of stuff in the next year, just in Los Angeles. And that was the beginning of a really big rush. *Switched-On Bach* came out a year later."

◯

How the Beatles Bought Their Synthesizer

One person who wanted a Moog synthesizer was George Harrison. Bernie Krause is still angry about what happened. The story begins in November 1968 with a late night session in an LA recording studio. The famous Beatle was working with a musician he was promoting, Jackie Lomax, and Bernie was doing his usual thing—adding some groovy synthesizer sounds. After the session finished (at 3 a.m.), George asked Bernie to stay behind and show him how to play synthesizer. Krause: "And I set up the instrument and was beginning to demonstrate things that Paul and I were considering for a new album . . . some patches that we were thinking of. And when I set these up I didn't realize . . . that he had asked the engineer to keep the recorder going . . . Harrison didn't ask my permission to do this, just did it, assumed that it was okay to do."

George ordered his own synthesizer, and eventually Bernie was flown over to London to show George how to play the instrument. Imagine his surprise when, on arriving at the Beatle's mansion, he found George had already composed a piece of music with the Moog:

> He had a tape recorder there . . . and I'm listening to this thing and I'm listening and I'm beginning to recognize parts of it as being the stuff I had done for the Jackie Lomax session.
>
> I said, "Harrison, this is my stuff." I said, "What is it doing here and why are you playing it for me?"
>
> He said, "Because I'm putting out an album of electronic music."
>
> . . . I said, "George, this is my stuff, we need to talk about how we're going to split this, how we're going to share this—if you want to put this out, I don't like it very much, but if you want to put it out, we've got to work something out."
>
> "When Ravi Shankar comes to my house he's humble," says George. He said, "I'll tell you what, if it makes any money I'll give you a couple of quid. Trust me, I'm a Beatle."

In the end, George Harrison's embarrassingly bad album, *Electronic Sounds* (1969), was released under his name alone on Apple's more experimental label, Zapple. A close inspection of the front cover (the best thing about the album)—a beautiful child's portrayal of a Moog synthesizer—reveals a silver line across the bottom under George Harrison's name. In certain lights you can read "Bernie Krause" under that silvered-out streak. On the inner sleeve, one side of the record is credited as being made "with the assistance of Bernie Krause."[14]

Recognition of their efforts was a problem facing all early synthesists. Was the actual creation of original electronic sounds—the patching or programming—an artistic or engineering achievement? With all its dials and wires, it was perhaps not surprising that producers and record-industry people regarded the Moogist as being more like a recording engineer. One can see the difficulty on other records where Beaver and Krause played. On Mort Garson's *The Wozard of IZ: An Electronic Odyssey* (1969)—an amusing psychedelic spoof on the movie—the credits acknowledge Bernie Krause as "the electronic producer." On *The Mason Williams Ear Show* (1968), the credits acknowledge that "Moog engineer Paul Beaver was in charge of plugging and unplugging." The record industry just did not know how to deal with this hybrid machine-instrument and its operators; it defied all the normal categories.[15]

The Beatles did go on to use George's Moog synthesizer on their acclaimed album *Abbey Road* (1969)—the last album they made together, and George Martin's favorite.[16] It was for many years the best-selling Beatles album. All the Beatles, apart from Ringo (who eventually bought and tried to learn on an EMS VCS3 synth), used the new instrument. The first track on which the Moog is used is "Because," where George Harrison plays three different Moog patches, including the emulation of a French horn sound with a sawtooth waveform.[17] John Lennon uses it on "I Want You," building up the sinister effect of this rocker with louder and louder white noise. On "Maxwell Silver's Hammer," the most elaborately produced song on the record and one that John Lennon hated, Paul McCartney uses the ribbon controller; in all, there are five different Moog parts on this song.[18] But the Moog *pièce de résistance* is without question one of the best known Beatles' songs ever, George Harrison's "Here

125

Comes the Sun." The Moog (played by George) can be heard throughout, its increasing brilliance of timbre reflecting the sun's increasing brilliance. The Moog company never referred to the Beatles' use of their synthesizer in promotional material, but nevertheless the Trumansburg workers were thrilled to have the Fab Four as their customers.

George Harrison recalled his experiences with the Moog: "It was enormous, with hundreds of jackplugs and two keyboards. But it was one thing having one, and another trying to make it work. There wasn't an instruction manual, and even if there had been it would probably have been a couple of thousand pages long. I don't think even Mr. Moog knew how to get music out of it; it was more of a technical thing. When you listen to the sounds on songs like 'Here Comes the Sun,' it does do some good things, but they're all very kind of infant sounds."[19]

For George Harrison, like so many rock musicians at the time, the Moog was a curiosity, a "technical thing" capable of "infant sounds" but little else. Perhaps George had forgotten that he had "wanted the Moog so bad" when hearing it the first time in the hands of Bernie Krause; certainly he was interested enough to release a whole album of Moog music. But without an instruction manual and without Krause to teach him how to play it, it became over time just another rich rock star's plaything.[20]

○

The Nonesuch Guide and Other Albums

Beaver and Krause's *The Nonesuch Guide to Electronic Music* is, as far as we can tell, the only technical manual in the history of the music business to become a hit record. The boxed two-album set was given the chic Nonesuch packaging and was accompanied by a 16-page booklet describing the basic elements of sound synthesis. Released in spring 1968, it spent 26 weeks in the *Billboard* Top 100—dramatic evidence for how the new craze of electronic music was taking off.

Part of the project was to develop a notation system for describing the dif-

ferent sorts of sounds the synthesizer made. Different symbols denoted different categories of sounds and the way they were produced. This ambitious notational scheme, although intended to be fully general, never really caught on because it was technology-specific and was made redundant when the new generation of portable keyboard synths like the Minimoog came along.

Less precise ways of describing the new sounds evolved in the studios. But there were so many different sounds that words seemed inadequate to the task. Some general terms applied like "fat sound" to describe the Moog filter, but the terms for specific sounds seem to have been largely idiosyncratic, like the elaborate patch diagrams each synthesist prepared.[21] One method was to try to capture in words the sort of sound involved. For example, Bob Margouleff and Malcolm Cecil used "trinkler" to describe an arpeggiator effect, "a sort of bright little star, twinkly little [thing]."[22] Another quirky way of describing a sound was David Borden's "building the house sound," named after the shape of the waveform producing the sound—a triangle on top of a square wave so it looks like a house!

Bernie found that the requested sounds often referred to a particular record they had worked on earlier. "Could you get us the sound you used on a Stevie Wonder date?" "The sound with the Byrds." After a time Beaver and Krause found themselves using a "limited repertoire of 20 or 30 sounds that we got, that were very easy to patch and do. Finally, it just got to the point where it was becoming so simple and ridiculous that we were able to replicate those sounds even on a Minimoog or a Model 10 [a smaller more portable modular system] and we didn't even bring the big synthesizer with us because nobody wanted to explore."

With their contacts in the music business and the growing popularity of the Moog, Beaver and Krause were finally able to secure a record contract with a major label, Warner Brothers. Their first record for Warner Brothers, *In a Wild Sanctuary* (1970), combined natural sounds from the environment with synthesized sounds and was a critical, if not a commercial, suc-

127

cess.[23] One of their new techniques, used on a track called "Spaced," was quickly copied by a famous Marin County film company and became a standard in movie theaters. It is played right at the start of the show: a distant note seems to get closer and closer to the listener before splitting into a tumultuous eight-note chord. According to synthesizer builder Tom Oberheim, the original analog form of the sound is much richer than the "digital perfectness" used in movie theaters.

Paul Beaver, the political conservative, was uncomfortable with the ecological theme and antiwar sentiments of *In a Wild Sanctuary*, and his convictions were further shaken by its critical acclaim. He squirmed even more when, shortly after its release, one of its tracks, "Walking Green Algae Blues," was adapted by Daniel Cohn-Bendit (Danny the Red) as the anthem for his new Green Party. According to Krause, Beaver found this "embarrassing" and also "appalling and amusing" at the same time. The political split in their partnership was summarized by Bernie as follows: "I marched and Paul played sessions." Paul had enough of a sense of humor to joke about their political differences and was not above pulling the occasional prank. Once they got even with a recalcitrant sponsor (ITT) who wanted a commercial but refused to pay the going commercial rate. ITT's logo was an old radio tower with a beacon on the top of it, "and we hit on an idea almost simultaneously to do Morse code, 'Fuck You' in Morse code on this tower, with the sound of the synthesizer in the background. And we did."

Gandharva (1971) was the last important Beaver and Krause album.[24] This is the first ever live quadrophonic recording and was made in San Francisco's Grace Cathedral, with its ninety-foot high nave and seven-second sound delay providing natural reverberation. It is one of Bernie's favorite recordings: "The effect certainly surpassed any chemical rapture I have ever experienced." The Grateful Dead sound crew brought in their sound system for use at the record-release party so that the full spatial grandeur of the work could be heard. Experiencing electronic sounds from a huge speaker system is one of the great audio experiences that the sixties created.

○

The Magic of Paul Beaver

Paul Beaver's death is shrouded in mystery. Bernie Krause remembers a chilling phone call he received from Paul a week before. Paul announced he was about to leave his body and told Bernie there was a complete inventory of their gear in his desk drawer and to make sure his precious pipe organ parts went to someone who could use them. Paul seems to have known that he was shortly to return to being a Thetan (the Church of Scientology's name for the inner being that inhabits the human body). People came to Paul's memorial service in droves to remember this shy and lonely man. There was no will, and the pipe organ parts, like everything else, sadly vanished as Beaver's family took what they could sell.

Paul Beaver, like almost all early users of the Moog, apart from Wendy Carlos, has been overlooked. He was everywhere and nowhere. He understood the technical capabilities of the Moog and had the keyboard technique to make the synthesizer come alive in a way that few others could at that time. Rachel Elkind, who worked with Carlos on *Switched-On Bach*, told us that they were anxious about only one set of competitors who might beat them to their spectacular success, and that was Paul Beaver and Bernie Krause. Tom Oberheim told us Paul Beaver was "other than Carlos, probably the person most responsible for getting the synthesizer thing going." As his friend and colleague Bernie Krause recognized all too well, "Paul had the magic. He was a magician, so they needed him."

Beaver and Krause, over time, became victims of their own success as more and more musicians took up the synthesizer and some started to play it seriously. After Paul's death, Krause began to lose his enthusiasm: "I went through a period of trying to figure out how to do stuff alone. By then, for me the synthesizer was becoming clichéd . . . And it wasn't getting to be very much fun anymore because all we did was replicate stuff . . . And digital was beginning to raise its ugly head."

Krause was witnessing the evolution of the synthesizer from an instrument that could produce a variety of unknown sounds to one that reproduced a standard package of familiar sounds. This process was facilitated by subsequent changes in synthesizer design. First by the Minimoog, which had switches and hardwired sounds rather then patches, and later by wave after wave of digital instruments. The paradox for Krause was that the very success of the synthesizer that he and Paul Beaver had worked so hard to achieve also sounded the death knell of synthesis as an exploration of sound.

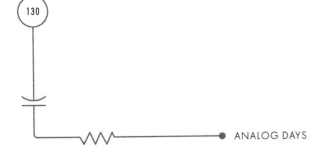

7

Switched-On Bach

> The whole record, in fact, is one of the most startling achievements
> of the recording industry in this generation and certainly one of the
> great feats in the history of "keyboard" performance.
>
> GLENN GOULD

IT WAS FALL 1968 and Jon Weiss and Bob Moog were attending a midtown record release party. The A&R people at Columbia Records were excited: they had three albums of groovy-new-weird-electronic kind of music on their hands. *Rock and Other Four Letter Words* (1968) was expected to do well. *In C* (1968) by Terry Riley might appeal to the avant-garde crowd from which it had emerged, and the heads were getting to love that sort of music. No one quite knew what the other album, *Switched-On Bach*, would do. It used the Moog and that had helped propel Beaver and Krause's *Nonesuch Guide* into the charts, but it was J. S. Bach and that made it, well, less than cool. As the mandatory bowl of joints circulated, the industry hacks, journalists, musicians, and hangers-on talked up the products. Bob and Jon watched bemused as Terry Riley, dressed all in white, got up and played the electric organ. There was no sight or sound of the unknown artist who had recorded Bach on the Moog.

Rock and Other Four Letter Words vanished without a trace.[1] Terry Riley's *In C* became a landmark record for minimalism, influencing rockers and composers alike, and did very well for Columbia. But *Switched-On Bach* changed the face of pop, rock, and classical music—the first classical recording *ever* to go Platinum. Reviewers predicted the Grammy-winning record would finally release electronic music from sounding like "some obnoxious mating of a catfight and a garbage compactor" and from its predictable use in "cheesy invader-from-Mars movies." Somehow its creator had managed to square the circle, producing electronic music that was dramatically innovative while at the same time being "music you could really listen to."

S-OB was a crossover album, appealing to pop, classical, and electronic music audiences. Its sensational debut embodied a little bit of a rush for everyone: it scared studio and orchestral musicians (and their union), who could see their jobs vanishing if just one synthesizer in a recording studio could now duplicate their efforts. It wearied experimental artists and avant-garde composers, who thought imitative synthesis was a poor use for the dazzling new technology. It inspired a plethora of inept imitators, who anticipated dollar signs. It delighted the public (and, naturally, the recording industry), who bought the albums as fast as they were put on the shelves. It made the Moog synthesizer famous—Moog and Carlos became overnight celebrities, immediately in demand for television talk shows, interviews, and personal appearances.

○

Ugly Atonal Styles

Born in 1939, Walter Carlos began piano lessons at the age of six and, with his parents' encouragement, continued to study classical music until he was fourteen. Interested in electronics at an early age, he won a scholarship at a Westinghouse Science Fair for projects involving computers, and by age fifteen was technically skilled enough to build a non-equal-tempered keyboard. At sixteen, he was altering his parents' piano to various "unorthodox tunings." By this point he had also become an accomplished organist.

A year later he became interested in electronic music, making his own tapes, mostly *musique concrète*, "plus whatever you could get out of a laboratory oscillator and splice into some kind of shape."[2] Throughout his college years at Brown University (1958–1962), Carlos pursued his twin interests in physics and music, eventually majoring in physics. This combination of subjects undoubtedly led to his expert technical knowledge of sound engineering: "I always want to peek and see how the magic trick is done . . . musicians are magicians. Our shop talk ought to be about how the illusion is produced with no holds barred."[3] He then moved to Columbia University, where he earned a master's degree in music composition working with Otto Luening and Vladimir Ussachevsky (1962–1965).

At Columbia, Carlos found himself in the middle of a debate that had been intensifying since the early part of the century. The deliberately atonal, highly systematized movement known as serialism clashed with the tradition of orchestrated tonal music, with its emphasis on melody, harmony, and counterpoint. Carlos hated serialism. "I didn't go for that type of non-rhythmic, non-melodic, non-harmonic music. It seemed more concerned with what we don't do than what we do."[4] The public too was unimpressed with serialism, preferring to attend concerts where more traditional musical fare was served up.

But in academic communities in the late fifties and early sixties, atonal music was dominant. At both Brown and Columbia Carlos encountered "alienation and condescension" on the part of both students and faculty toward his own traditional musical values. With his background in science, he was also skeptical of serialism's mathematical pretensions: "That kept me out of peer groups of students who . . . all got into serial mathematics and 12-tone rows. Having a math background, I thought that it was all gibberish."[5] He quickly learned to keep quiet and discovered that his appreciation for tonal music meant that at college he was not thought of as "a composer in the wider sense."

◎

Electronic Music as Sanctuary

Realizing that his chances of making it as a composer were limited in the prevailing academic climate, he switched his emphasis back to his childhood hobby, electronic music. But here too he found that the academic avant-garde had left their mark: "The general public considered it to be avant-garde in the worst sense, completely without redeeming value or commercial interest." Carlos resolved that what the new field needed was an old-style touch: "I thought that if I offered people a little bit of traditional music, and they could clearly hear the melody, harmony, rhythm and all the older values, they'd finally see that this was really a pretty neat new medium."

His background, combining technical expertise with composition, was an ideal preparation for his chosen field. For Carlos, electronic sound became a kind of sanctuary—an escape from all that he hated about the academic world of composition. Ussachevsky, the Director of the Columbia-Princeton studio, was a welcoming presence, allowing him free use of the equipment provided he worked the night shift (usually from midnight until dawn). Phillip Ramey, a fellow graduate student in musical composition, describes what those times were like:

> One of my most vivid recollections of those years is of countless night sessions in the Electronic Music Center, and of Walter and myself emerging onto the campus in the early morning, blinking dazedly in the sunlight and staggering across Broadway to the local Chock Full o' Nuts for coffee . . . Walter was maniacally involved with the tape machines . . . each of us worked in constant dread of marauding janitors who seemed unaware of the Ussachevsky-Carlos compact.[6]

Carlos's intensity and his commitment to the new field singled him out, but he had yet to acquire the instrument and explore the genre of electronic music that would make him famous. His Columbia compositions, such as his very first "Dialogues for Piano and Two Loudspeakers" (1963), were clearly influenced by the experimental genre of electronic music and were still a long way away from the "melody, harmony, rhythm and all the older values" that Walter hoped would be the redeeming feature of the new medium. Indeed it seems that Carlos hoped that the more popular pieces, realized later, would help gain acceptance for his more "adventurous" works.

○

A Vocabulary that Spoke Telegraphically

After graduating from Columbia, Walter continued to hone his technical skills, working for three years as a recording engineer, tape editor, and disc cutter at Gotham Recording Studios in mid-Manhattan. He knew that he wanted a synthesizer, and as he looked around he was drawn to the one man in the vicinity who could provide it. He had first met Bob Moog at the 1964 AES convention. He bought a small Moog synthesizer in 1966 that Bob delivered in person to his modest rented apartment on West End Avenue (staying the weekend to make sure it worked properly). Gotham eventually allowed him to haul his newly acquired synth into the studio and store it there, enabling him to use their superior studio equipment for recording, overdubbing, and mixing.

As word spread among the studio's regular clients about the existence of Carlos and his synth, a few much-appreciated jobs followed, such as commercials for Schaefer Beer and the Yellow Pages. Jon Weiss heard those early commercials, and already it was obvious to him that Carlos had a technical skill beyond that of most synthesists. But given the time-consuming nature of putting together electronic music, the temporary arrangement with Gotham was not satisfactory. To progress, Walter realized he would have to assemble his own electronic studio: "Somehow I needed to put together a minimum configuration of a multitrack machine, a two-track stereo machine for mixes, and a basic console/mixer with monitoring and other usual functions."[7]

As luck would have it, another Gotham studio engineer, Bob Schwartz, with expertise in the design and maintenance of studio equipment, became

135

intrigued by Walter's dream. After finishing work for the day at Gotham, they sketched out plans for the studio, located the necessary pieces of hardware, and assembled it all in Walter's apartment. By the time they had finished, Schwartz and Carlos had become firm friends and the ground had been laid for the real work to begin.

Carlos was not one to be satisfied with just any old synthesizer. He began conferring with Bob to discuss additions and improvements. Reynold Weidenaar remembers Carlos as one of the musicians who worked closest with Bob: "Carlos was very clear [about what he wanted], and I remember there was some frustration because he was really holding Moog's feet to the fire in terms of the way things had to be, and the quality that he needed . . . a very demanding musician who's also very knowledgeable technically . . . This is what he had in Carlos, and he valued that highly."

Carlos's input was very specific. Soon after he purchased his Moog, he realized he had a need for a portamento and hold switch "to delay the intervals between each 1/12-volt step" on the keyboard. Carlos: "Before the hold switch was put on, if you took your hand off the keyboard the frequency went to 12Hz or something. It was terrible until Bob came up with that. That was back when you really felt like you were working with an invention. I miss that time. Bob Moog has a wonderful feeling about music. It was perfect for me because it's hard for me to talk about things, and between the two of us there was a vocabulary that spoke telegraphically."[8]

Bob's detailed knowledge of electronic musical instruments and Carlos's own increasingly refined sense of what he wanted from the new medium made a perfect pair. Carlos, by reputation, was shy and intense, and this melded well with Bob's own slightly unworldly personality. Carlos was also instrumental in helping Bob tweak his design for a touch-sensitive keyboard into a workable mechanism. Bob fully recognizes Carlos's input into his project: "Yeah, and he—she—uh, was always, you know, criticizing—constructively criticizing—telling me what kind of knobs feel good and things about the sound, what kind of function she wanted."

Carlos had his own reasons to help Bob. "Everything going out to Carlos was custom," according to Weidenaar, with "much higher specifications than the standard modules." Weidenaar does not remember seeing Carlos in Trumansburg, and if it occurred "it didn't happen often." This was because Bob often traveled to New York to visit his parents, at the same time bringing prototypes back and forth. Although a central figure in the early history of the Moog, as an individual, Carlos, Castaneda-like, was a shadow, a recluse, quiet and mysterious. Many people who were around at that time, like David Borden, knew of him, and his importance to Bob's project but did not know Carlos as a person. "I never knew Walter— Wendy. I'd hear about, and I remember hearing about the operation and everything, from Bob . . . and he was very close."

○

Personal Empowerment

The gender ambiguity in Bob and David's recollections of Carlos is explained by the fact that, at precisely this time in his life, "he" was becoming "she"; that is, Walter was changing to Wendy. Walter began hormone treatments and cross-dressing early in 1968 and "permanently [living] as a woman in the middle of May 1969, nearly three and a half years before the [transsexual] operation" in the fall of 1972.[9]

As one of the very first public figures to undergo such a change, Carlos was to be a pioneer in more ways than one. As Rachel Elkind, Wendy's friend and collaborator, told us, "You have to remember this was 1968, there was one transsexual in the whole world that anybody had heard of. That was Christine Jorgensen." The question arises as to whether Wendy's metamorphosis, which occurred just around the time she was developing as a synthesist, had anything to do with the Moog, and with synthesis itself. Perhaps there was something about this most unusual instrument that resonated with the most unusual transformation its star performer was about to undertake.

The question of gender and the synthesizer is a tricky one. Certainly electrical music technologies have traditionally been used for building masculine identities—the boys and their latest toys. But different sorts of masculinity can be involved in how men interact with technologies, and several women we interviewed for this book, notably Suzanne Ciani and Linda Fisher, have developed intense personal relationships with their synthesizers, as we will see. If, as Judith Butler argues, gender identities have to be performed, a key prop in the performance of these synthesists is the machine with which they spent most of their waking hours interacting—the synthesizer.[10] What we want to suggest with Wendy and her synthesizer is that it may have helped provide a means whereby she could escape the gender identity society had given her. Part of her new identity became bound up with the machine. The transformative power of the synthesizer may have allowed her not only to conjure up a new musical meaning but also helped her find herself as a newly gendered person. While some people used the transformative power of the synthesizer to escape from the prison of "straight" society, to help them transcend to new states of consciousness, Wendy, we suggest, may have used it to help her transcend her former body and her former gender identity.

○

Transcending the Limitations

Wendy was not alone in her work. At Gotham Studios she met her future collaborator, producer Rachel Elkind. With a background in jazz and musical comedy, Rachel had come to New York to work on Broadway. She ended up getting a PhD in music and working for Goddard Lieberson in the recording industry. According to Wendy, their initial meeting was "loathe at first sight. We didn't care for each other at all. It took us about a year before I started bugging her to collaborate with me or produce me."[11] Recognizing that Rachel, with her knowledge of the recording industry, was someone who could help her, Wendy brought her some of her early arrangements on the Moog; the first piece Rachel heard was a synthesized version of "What's New Pussycat." Rachel was unimpressed. It was not until she heard Wendy's version of Bach's "Two-Part Invention in F Major" that she realized what Carlos had stumbled upon: "I really felt that that was something that could really speak, that transcended the limitations of the instrument." Bob Moog also heard that first piece and points out that it is played too fast, done before Wendy had really got the hang of how to do it right. But that early piece had something Rachel and Bob both recognized.

As well as Bach, Wendy was also experimenting with rock pieces, making commercials, and continuing to work on her own original compositions. It was Rachel's idea to do a whole album of Bach. "And she said, 'A whole album of Bach?' And I said, 'Yes, I think so,' because my thing was music had to sing and dance and had to have truth, and if it did, then it would speak to an audience." Wendy's own compositions were still far more experimental pieces than the known and chartered territory of Bach.[12] But Rachel, savvy to the recording industry, recognized that an unfamiliar instrument with an unfamiliar composition was not an alliance ticketed for success. Wendy remembers, "People couldn't even pronounce" the word "synthesizer." It was so unfamiliar that when they were working on S-OB, "some of the producers didn't want us to use the word."[13]

Rachel's conclusion was that they could not find a better composer than Bach. And the prospects for a hit were not unprecedented. A London rock band, the Nice, starring Keith Emerson, had scored a surprise hit in Britain with their rock version of the Brandenburg Concertos.[14] The counterculture was also not adverse to a bit of Bach. The organist Virgil Fox played Bach at sell-out concerts he performed at the Fillmore West. The venerable composer had been reworked in many mediums and was ideal for yet another outing.

Having decided to create an album of Bach together, Wendy and Rachel began working on the first movement of the *Third Brandenburg Concerto*. Wendy did all the synthesizer parts and Rachel produced the album. The

139

musicologist Benjamin Folkman, a friend of Wendy's, also contributed "proper performance practice and idiomatic Baroque ornamentation." Because Benjie was involved, Rachel was confident the results would "be really, really terrific and very salable." Folkman had established credibility in the music world, and his opinion counted. Rachel: "That was important because it sort of allowed the work to stand without tremendous criticism from the classical press. Even if there was, they had to accept that it was really authentic and interesting in its own way."

Wendy and Rachel's working styles were very different. Wendy has an obsessive personality, knowing how each note was realized. Rachel's style is more intuitive, with an emphasis on improvisation. Rachel: "At that time the studio was in her apartment. And [I] would go over and we'd work together. Putting together electronic music was a very tedious process . . . I really came from a very improvisatory discipline. And I think that really worked because I think I pushed it to become as alive as possible, and I think that's what distinguished our music from a lot of the other electronic music that came after it." With the need for endless overdubs and for layering the music, the sound got "thicker" and "you'd have to lighten it up and maybe change the timbre here or there. And that's how I became the critical ear."

As the album unfolded and Rachel realized that Carlos was in the process of achieving a breakthrough, she became nervous about being scooped by the likes of Beaver and Krause. In the end, Rachel believes that Wendy, with classical training and an attraction to polyrhythms, may have been the best-suited to bring the project to fruition.

For Wendy and Rachel, S-OB was not a didactic exercise. They wanted to make the music come alive, and finding new timbres was an important part of the process. They were on the frontier in trying to coax a whole new range of sounds out of a cluster of electronic circuitry. To this extent, S-OB was not imitative synthesis, although as they worked side by side they relied on the language of contemporary musical idioms to reach for the sounds they were after. Rachel helped Wendy search for certain timbres;

she would not want to imitate a French horn exactly but she might suggest that Wendy craft the sound to be "a little like a French horn, a little more mellow." Indeed, this is part of the achievement of the record: the timbres sound familiar, yet they are clearly new and different electronic timbres.

By Wendy's own estimate, there were only about "half a dozen basic sounds" available in an analog instrument. So how could she possibly manufacture the varied tonal nuances of *S-OB*? Bypassing the Moog's voice limitations, Wendy developed her rich musical range by learning how to rapidly "jump from timbre to timbre," so that, according to Wendy, listeners (including Bob) imagined they heard "greater timbral resources than really existed" in the machine itself.

Wendy was the ultimate technical craftsperson; her technical proficiency on the instrument was unsurpassed. Bob has a lot of respect for Wendy, as well as fondness and a continuing friendship: "Wendy used techniques that had been available for years—but used them better."[15] Rachel will go one step further: "She knew it better than Bob Moog," and "just was one with that instrument."

Wendy's results are partly attributable to the fact that she was an experienced sound engineer, with excellent splicing, over-dubbing, and recording skills. Reynold Weidenaar remembers visiting her and hearing an early track: "Carlos played some of it for me at her apartment, the original master. I can see the splices go by, see the edits, see the timbral changes. Every time you saw a splice coming up you'd know that you were going to get a different voice. So I saw the bits and pieces, as it were."

○

Bach-to-Rock

It took the spring and summer of 1968 to complete *S-OB*. With the music going well, Rachel plotted how to get a record contract. The music business was still "very much a man's world," so she persuaded a colleague and friend, Ettore Stratta, in A&R (Artists and Repertoire) at Columbia to submit the proposal for her. This worked. They were offered a thousand dollars for the finished master (this was half the advance Beaver and Krause received for their *Nonesuch Guide*), and a two-album commitment from Columbia. They were thrilled, although it was clear to both of them that the record company had no real interest in them personally. It just so happened that *S-OB* was a fit with Columbia's marketing scheme, which at the time was a Bach-to-Rock campaign.

Jon Berg, the art director at Columbia, came up with what, in hindsight, was a stroke of genius—the album's title. Wendy and Rachel had been toying with various catch phrases such as "Electronic Bach," but it was Jon who hit on the prefix "Switched-On." Rachel conceded: "The minute I heard it I hated it, but I knew it was the right title, you know?" "Switched-On" conveyed perfectly the electrical origins of the sound, plus the appeal of being tuned in and turned on. The cover photo—a wigged baroque musician quizzically listening to a Moog synthesizer—continued the symbolism. The keyboard in front of the panel of knobs tells you that this is an instrument, and the power cords let you know electricity is involved—but laughably, for those who knew anything about the Moog, there are no patch cords. This synth would have been unable to utter a bleep, never mind play Bach.

Columbia's interest in Rachel and Wendy was no doubt in part a response to the pressure they were under from Jac Holzman's newly developed Nonesuch label. Jon Berg's catchy cover design owed much to other baroque albums being marketed by Nonesuch in a thoroughly modern way. Nonesuch's *The Baroque Beatles Book* (1965) juxtaposed wigged baroque musicians with Beatles songs, including one wearing an "I like the Beatles" tee shirt.

○

Silver Apples

Just before *S-OB* came out, another important electronic music composition appeared, more in the style of experimental music but nevertheless appealing enough to produce significant classical sales and to even become an underground hit. Morton Subotnick's *Silver Apples of the Moon* (1967) was commissioned by Nonesuch, and the entire piece was made on a Buchla 100 synthesizer. Subotnick by this point was also in New York, having had Buchla build him a replica of the original Buchla Box 100 before leaving the Tape Center. The exciting tonal colors, the spatial movement of the music, the rich counterpoint of gestures, and the purity of the sound (particularly the sine waves) were all elements that typified the Buchla. As one reviewer noted, "It's a beautiful record . . . it seems to glitter with precision."[16] The sequencer-generated rhythmic sounds soon found a home in dance and ballet.

Interestingly, given that her own record was due out soon, Wendy reviewed—and panned—Subotnick's record for Bob's magazine, *Electronic Music Review*. Although conceding it to be "one of the 'prettiest' electronic compositions" that had been released up until that time, she had to admit, "I'm sorry, but 'Silver Apples' is a bore." She complains that perhaps the album is too long "for a single electronic composition of this style and type"; or the problem might be with the Buchla itself, which "contains certain operational 'traps' [such as the sequencer] which are avoided only with great difficulty." Wendy's most damning criticism was that "the phrasings and articulations are not particularly expressive; they either sound inflexible and mechanical, or aleatoric and unimportant." In summary, "All is euphoric and pleasant, but never musically compelling." Wendy does cut the composer himself some slack at the end of the review, declaring that although "Silver Apples" turns out to be "a poor performance of a fine composition," it and "the very talented Morton Subotnick" are to be commended.[17] Wendy's criticism of the record seems to be as much about criticizing the Buchla as about criticizing Subotnick.

A comparison between the two most famous works on the two synthesizer pioneer's different machines—the Moog and the Buchla—reveals a paradox. The Buchla had been designed to make music in real time and as

143

an instrument that the performer could really interact with. Wendy, on the other hand, with the Moog had had to use endless tape dubs to produce her masterwork. Yet it was Wendy's music, which could never be performed, that was the more expressive, the more alive, the more like a performance. Although Subotnick's record won critical acclaim and sold considerable numbers for that sort of experimental record, it was Wendy's record that achieved the breakthrough.

○

Something Wasn't Right

Always reclusive, during the production and then the release of S-OB Wendy did not make many public appearances. An exception was her presence (as him) in the audience for Bob's famous October 1968 airing of a selection from the album at the New York AES meeting (just before the album was released). It was part of a paper Bob presented, "Recent Trends in Electronic Music Studio Design," and Bob played Carlos's realization of the *Third Brandenburg Concerto*. The story of its impact is one Bob loves to tell: "I put the tape on, and I wanted to let it run. So I just walked off the stage into the back of the room. And I can remember peoples' mouths dropping open. I swear I could see a couple of those cynical old bastards starting to cry. At the end, she got a standing ovation, you know, those cynical, experienced New York engineers had had their minds blown."

The success and instant notoriety of S-OB, where synthesized sound was finally "acclaimed as real music," demonstrated that the medium could be used for electronic music the public could appreciate.[18] Wendy and Rachel's achievement was a success beyond their wildest dreams. But there was one down side—the album thrust Wendy unexpectedly into the limelight, just as she was trying to keep a low profile in order to undergo her transformation. Although Wendy had thought of herself as a woman from well before S-OB, her public persona was still Walter. Wendy's transformation certainly overwhelmed her ability to perform in public and interact with other musicians and listeners. It was for her a "very sad time."[19] She had success but she couldn't enjoy it.

Rachel, who by this time was sharing her West 87th Street Brownstone with Wendy, was a witness to the pressure she was under. Living together was partly an effort to protect Wendy during her metamorphosis. Rachel herself recalls being "so neurotic that people were going to find out about Wendy's situation." During this period, Wendy would not appear in public, and she felt that she had to hide herself from other musicians. When George Harrison or Keith Emerson appeared at the door, Wendy would listen from "upstairs" as Rachel explained to them that "Walter was away."[20] When Stevie Wonder came over once to play the synthesizer, realizing that he had exquisite hearing, she did not speak to him for fear that he would hear her voice and realize that "something wasn't right": "The fact that I couldn't perform publicly stifled me. I lost a decade as an artist. I was unable to communicate with other musicians. There was no feedback. I would have loved to have gone onstage playing electronic-music concerts, as well as writing for more conventional media, such as the orchestra."[21]

She found herself becoming a star but was unable to make live appearances. As time went on, the folks at Columbia became "disinterested" in Carlos; they needed to showcase "a real artist" that "they could have in pictures and stuff, and running around concertizing."[22] The personal issues must have been agonizing. For a 1970 appearance on the Dick Cavett Show, Carlos dressed as Walter, and in ads from this period, for instance, standing in front of a synthesizer advertising the Dolby Sound System, Carlos is dressed as a male with prominent black sideburns.[23]

In 1969 after her follow-up album the *The Well-Tempered Synthesizer* (1969), Wendy made one concert appearance with the St. Louis Orchestra. For Rachel "it was just such a nightmare," that she decided enough was enough; she told Wendy "it wasn't worth it and I would never sort of do it again." Just before the show, Wendy "began to cry hysterically" and informed Rachel that she did not want to proceed with the performance. She

145

had arrived at the theater dressed in women's clothing, but now the necessity of getting up in front of all those people as Walter, was, understandably, overwhelming. In what must have been a desperate the show-must-go-on spirit, Carlos "touched up his face, which the estrogen had softened. He pasted on sideburns, stuffed his long hair under a man's wig, ran an eyebrow pencil over his smooth chin to simulate 5 o'clock shadow," and went on with the concert. After this experience "Walter Carlos refused to perform in public again."[24]

When their collaboration began, Rachel did not know that Carlos had a hidden personal issue that would impact their work and how it might be received by the public: "At the time that I was working with Wendy I did not know about her gender problems. In other words, I sort of accepted her just as she was, a wonderful human being. And it was really after I had made the deal with Columbia that she told me about this problem, which is why the album cover really was done the way it was with 'Trans-Electronic,' because she really didn't want to have a name like the Beatles or the Rolling Stones."

Walter, envisaging a new persona, was already trying to make space for Wendy. Rachel, too, was facing a difficult time adjusting to their unaccustomed success. As well as managing the complications of cross-gender politics, there was also the thorny issue of her own role in the partnership. "Having built up Walter Carlos, I also got tired of people thinking that I was there serving tea." It is not surprising that they both felt as though they were living two lives during this time. Rachel: "Truthfully, I was juggling many things because I was not only producing the record, I was acting as the lawyer-negotiator, and I was also protecting sort of Wendy's persona . . . [we were] living a hidden life, as it were. So I know I felt a lot of pain that we couldn't sort of celebrate this and be really out with it . . . the thing is that I think we both felt that we didn't want it to become a circus."

Even if Wendy had been able to accommodate the requests to appear in public with her synthesizer, the Moog was almost impossible to play live, and certainly in a way that reflected her artistry. Wendy's great achievement had inadvertently led her into this conflict: *S-OB* was totally a studio production. Wendy was not only trapped by her gender but by her own proficiency and meticulousness. She was in the same situation as the Beatles after *Sergeant Pepper*—a milestone production for its unique sound but an effort that could not be reproduced live.

○

Carlos's Achievement

Regardless of the mixed evaluations *S-OB* received, depending on which side of the popular/avant-garde or traditional/modern divide the critic resided, there is no doubt that Wendy changed the public's notions about electronic music and the synthesizer. Everyone we have talked to for this book, even people in very different musical genres, freely acknowledges the impact of *S-OB*. For such notable keyboardists as Keith Emerson, Patrick Gleeson, Tomita, and Stevie Wonder, it was *S-OB* that switched on their own interest in the synthesizer.[25] As the years passed and the *S-OB* imitators multiplied, the singularity of Carlos's achievement has become more and more apparent.

Wendy and Rachel went on to many other projects, including three other Switched-on-Bach-like albums: *The Well-Tempered Synthesizer* (1969), *Switched-On Bach II* (1973), and *Switched-On Brandenburgs* (1979). They composed the scores for two Kubrick films, *A Clockwork Orange* (1972) and—their last work together—*The Shining* (1980), a horror movie. These classic films continue to be watched and talked about. The surreal tensions and eerie ambiance that each score provides has much to do with the movies' impact. Synthesizer sounds had finally come of age.

At this point (in 1980) Rachel got married, having met her husband (an astrophysicist) on one of the eclipse-chasing trips that were among Rachel

147

and Wendy's few indulgences (Wendy is fascinated by astronomy). When Wendy herself found a new companion, Rachel felt that it was time for her to move on.

○

Union Troubles

S-OB had a dramatic impact on the entire music and recording industry. One effect was totally unexpected. The Moog synthesizer was for a time banned from use in commercial work. This restriction first surfaced in a contract negotiated between the American Federation of Musicians (AFM) and advertising agencies and producers in New York City in 1969. The union was worried that following on from Carlos's success, the synthesizer was going to replace musicians. Indeed, this possibility was noticed before S-OB came out, when Rachel Elkind played it for the famous jazz bassist Ray Brown: "It was really important to me to have jazz musicians appreciate this. And he told me that this was going to be very bad for musicians, and I said, 'No way, how can you say that? They'll never replace the richness of a real instrument, this isn't as fabulous, blah, blah, blah.' But it turned out that his fears were correct."

While recognizing its potential for emulating other instruments, the crucial point for synthesists was that the Moog should be treated like any other instrument, and playing it was anything but easy, a point the union had yet to grasp. Moog: "Basically the union didn't understand what the synthesizer was. They thought it was something like a super Mellotron. All the sounds that musicians could make somehow existed in the Moog—all you had to do was push a button that said 'Jascha Heifetz' and out would come the most fantastic violin player!"[26]

Out on the West Coast, Paul Beaver and Bernie Krause were running into similar difficulties with the union. Bernie Krause: "The AFM threatened to shut us down unless we promised never again to try and emulate strings and/or horn sounds, thereby replacing other musicians."[27] The battle with the union was eventually won by the combined efforts of Paul Beaver and Walter Sear. Beaver found that the union had lost an earlier restraint-of-trade case when they had tried to prevent a rhythm machine being used to accompany a Hammond organ. Beaver threatened to take the union to court based on the precedent of this earlier case. Meanwhile in New York, Walter Sear, an old union hand, managed to convince them that the Moog was little different from a Hammond organ and still required a skilled musician to play it. The category of "synthesizer player" was eventually accepted into the union, although synthesists still experienced suspicion and hostility from the union well into the 1970s.

Several synthesists have pointed out to us that, indeed, the union's fears were well grounded. As Suzanne Ciani, who worked in the New York studios and saw its impact directly, told us, "Actually over the years the impact of electronic music in studio production in New York was drastic." Almost a whole generation of session musicians were put out of work by the synthesizer. On the other hand, there is no doubt that the growth of the synthesizer industry and the new sorts of musician it encouraged led to plenty of new work. The success of the synthesizer, without question, in the long term led to a major change in the business, to be ranked alongside earlier upheavals, such as the one brought about when the talkies replaced silent movies and the live musicians that accompanied them were put out of work.[28]

It Became like a Factory for Awhile

Bob Moog likes to joke that Wendy was the first person to make real music on the synthesizer. "You know what real music is for the record industry? Music that makes real money!" Everyone knew that Wendy had sold a ton of records. There were dollar signs in the electrified air. With commercial music producers believing "it couldn't be the artist—it had to be the machine," the switched-on copycat industry was born. It resulted in literally

Figure 20. Switched-On Santa

hundreds of albums being rushed out that all used the Moog in some way, shape, or form. Musicians and recording industry hopefuls wanted part of the Moog action in order to replicate Carlos's success. With names like *Switched-On Bacharach* (1969), *Switched-On Rock* (1969), *Switched-On Nashville Country Moog* (1970), *Switched-On Gershwin* (1970), *Switched-*

On Santa (1970), *Chopin á la Moog* (1970), *Moog Power* (1969), *Moog Espāna* (1969), *Moog Plays the Beatles* (1970), and *The Plastic Cow Goes MOOOOOOG* (1969), it seemed as if every corny title and genre of music was ripe for exploitation.

These pseudo-Moogists needed to produce fast, while the public was still attentive. Very few of these composers, arrangers, and performers approached the task with anywhere near Carlos's artistry (one exception was Dick Hyman), and none were anywhere near as successful. Jon Weiss, who personally demonstrated Moogs to

Figure 21. Switched-On Gershwin

some of these new visitors, quickly found that all they wanted was a cash cow that went "Moog": "I saw this influx of the most disgusting, copycat efforts . . . Some of the most insipid garbage."

Bob remembers well a recording session for *Moog Espāna*:

> We got a call from RCA, you know, would we help them? Next thing there's a pickup truck with an eight-track recorder on it, came up from New York City, pulled up at our door, they unloaded this eight-track recorder, which is like . . . a supercomputer is today . . . These guys came in with their cigars and, "Gimme something" [imitates speaking with a cigar in mouth] like this, you know, a New York redneck. An entertainment business redneck, you know? They're very crass, and their cigars are very smelly, and I asked Jon to do this, go into the studio and do this. That poor guy

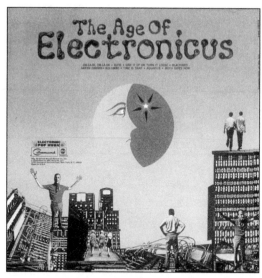

Figure 22. The Age of Electronicus

was in there all day, and he was shell-shocked. Here's this sensitive, artistic guy, and it was — musically, it was not — it was dreadful.

Another, more welcome impact of *S-OB* was that, for a short while, the Moog company (newly incorporated in 1968 as R. A. Moog, Inc.) could not keep up with the orders. Bob:

Before *Switched-On Bach* came out, and a couple other things, nobody believed that this kind of thing could be used for anything more than a novelty. You couldn't make real music with it, you couldn't be expressive with it. You couldn't make it swing. Then Carlos and a few other people demonstrated they were wrong. You know, they just [made an] end-run around the music business. And then, you know, in 1969, all hell broke loose. Everybody had to have, you know, every commercial musician had to have a synthesizer. Well, [we] had to hire people and buy parts.

Jon tells the same story from his perspective: "The difference in the Moog Company was astronomical. Before *Switched-On Bach* it was a lazy, sort of experimental concept that we're making this machine that some universities would use. And then after there was this explosion of interest, and he hired a business manager and new staff, the production went way up and they were testing things around the clock. It became like a factory for awhile."

One last impact was that Trumansburg was suddenly a destination, if not

Figure 23. Music to Moog By

a national musical landmark. Many more musicians, composers, and commercial sound engineers began visiting the Moog factory. Borden: "A lotta jazz guys came through just to look at it, and people who were doing electronic music before also came through to check it out . . . I remember being surprised that J. J. Johnson . . . one of the great trombone players, he came by, and we thought he was using it for a jazz instrument, but no, he was using it for commercials."

Unquestionably, 1968 and 1969 were boom years for R. A. Moog, Inc. Carlos's big hit coincides with this period; and, looking back from thirty years later, it looks like this hit came out of the blue and led to the Moog's success. The real story, as we have seen, is much more complicated. Bob was preparing the groundwork for years beforehand. Synthesized sounds had already been introduced to the public by commercial musicians like Eric Siday (and even Carlos), who used them for sound effects, logos, and signatures. "Good Vibrations" (1966), although using a modified theremin and not a synthesizer, popularly connected far-out, electronic sounds with rock 'n' roll. And when Bob Moog offhandedly says that there were "a couple other things" besides S-OB that led to the interest in synthesizers, it is understating the impact of what was building toward an influential trend. The Moog was featured in a number of pre-S-OB albums and some of the best known rock groups were using the Moog a year before Carlos's hit, as a result of Beaver and Krause's successes on the West Coast, particularly after Monterey. And their success in turn built upon the psychedelic movement that had emerged a year earlier from the Trips Festival and the influence of Kesey, Sender, and Buchla.

S-OB was in reality part of a much wider cultural transition encompassing the changing expectations of musicians and listeners—electronic sounds were now in the culture. S-OB was in effect a Trojan horse. Bach had, as one reviewer noted, been made psychedelic, but it was still Bach—and the synthesizer had been snuck in with it. Perhaps in the long term S-OB's impact as a musical achievement will be seen as an oddity, a footnote. But in terms of the history of the synthesizer and popular culture, Carlos's influence was unsurpassed. It brought the synthesizer from psychedelic obscurity fully into the mainstream, where it has remained ever since.

8

In Love with a Machine

And the fact that it's all this goddamn hardware, you know, that made it a guy thing too. It was halfway between being a musician and hot-rodding your car.

Bob Moog

SUZANNE CIANI was the first woman to make a name for herself by composing commercial sound signatures. She was known as "the woman who could make any sound." From the radical world of countercultural Berkeley to New York City corporate life, she was accompanied by a machine that, to her, "was my life. I mean, it was, I was in love." That love was for a Buchla 200 synthesizer.

Suzanne grew up just outside of Boston. She knew from very early on that she wanted to become a composer, gravitating toward her parents' Steinway piano and playing her sister's piano lessons for her when she was "very, very little." She studied music composition at Wellesley College and got her first taste of electronic music when her class visited MIT, "and the professor there was trying to get his computer, which at that time was this enormous thing, to sound like a musical instrument. And it emitted like one little beep, and this was like the hope of the future."

It wasn't much, but it did give her the idea that machines could make music. In 1968, having completed her degree, she enrolled in a two-year master's degree program in composition at Berkeley. She immediately became caught up in "the whole hippie thing . . . I mean I never wore shoes, hair down to my waist, we ate soybeans and brown rice, I hitchhiked to school. It was very counterculture." The effect on Suzanne was dramatic: "I found myself in the middle of complete revolution. You know, so I'm in the music building playing Chopin, a rock comes through the window—literally—and suddenly life was never the same."

This time of political turmoil was also a time of musical turmoil. She found, as Wendy Carlos had, that the academy was obsessed with serialism, which for her "had nothing to do with emotion." She began to explore other options. She heard that a center at Mills College had a couple of synthesizers that were not being used. Many nights she was up till dawn learning how to use the Mills College Buchla. Her first piece, "Breathing," for her Berkeley class had to be played in the campus theatre because there was no suitable tape deck in the music building: "It was just a sustained tone that shifted. And what you listened for was the evolution in the filter and the music that happened in the overtones—there weren't any notes. And so to me this was just sheer beauty, to hear this kind of delicate motion, and I was very proud. And after I played it, you know, [the] professor . . . said, 'Just tell me one thing,' and I was so excited, and I said, '*Yes?*' He said, 'Why did you bother to bring us all over here?'"

Suzanne's developing electronic minimalism was clearly too radical for her professors. Her frustrations continued. The Berkeley Music Department acquired a Moog synthesizer, but she was not allowed to use it until she had taken a course in synthesis, even though she was already composing her own synthesizer music. Eventually, she took a class with Bernie Krause and became an authorized Moog user.

Suzanne was fascinated by the Buchla and wanted to meet the man who had invented it. Through the artist Harold Paris (whose studio in Oakland was near Buchla's) she visited Don's workshop and got herself a job. The people who worked for Buchla were part of a new era, and most of them weren't concerned with ambition or money. During the workday the radio was tuned to KPFA, the Berkeley alternative station. The soldering tables were lined with workers you wouldn't typically envision, like "philosophers, poets, dancers, and Sanskrit specialists!"[1] They weren't trained in electronics, but they wanted to be involved in a business that was opposed to "the system."

Buchla ran a strange shop; during soldering, "no one was allowed to talk," presumably for fear that they would be distracted from the task at hand. Suzanne: "It was a large warehouse, and our assembly room was in one room. Then there was a very large open space which was his private studio, which was very dark, and you could walk in there any time of day and the Buchla, the machine, would be on, so thousands of lights running around. There was a swing attached to the ceiling. Well, the swing was to be able to relax while you were thinking."

Working in Buchla's shop was a very different experience from Bob's funky factory. In Trumansburg, one had the impression of an ordinary American small business—an assembly line in a rural town, with local workers and "easy listening" tuned in on the radio. The shop environment was working-class and matter-of-fact. Buchla's off-beat shop—from the lighting, to the piped-in politics, to the ambient countercultural atmosphere—was more like an on-the-job "happening."

Suzanne spent hours using Don's studio system, having won herself special privileges there. Slowly she mastered the intricacies of the Buchla, in between relaxing on the swing.

○

"I Could Run the World If I Wanted"

Keeping the job that Suzanne had coveted was no easy task: "I practically begged him to work there. And after the first day . . . at the inspection time at the end of the day Don found a cold solder joint, and he said, 'Well, it must be the new girl,' and he fired me . . . I said, 'It's not mine, I didn't do it, I *did not* make that cold solder joint and you cannot fire me' . . . So the next day I showed up at work. I said, 'You can't fire me, I'm here,' and so I continued."

IN LOVE WITH A MACHINE

Her boss, although a countercultural guru, was still a fifties man: "We had a lot of confrontation also because we were from different generations . . . I was a liberated, new woman. I could run the world if I wanted. And Don came from a generation where women were appendages." Suzanne asked Don if he would give a class on electronics, Don agreed, but Suzanne found herself the only woman there. "And after one day of class he said, 'I'm sorry, but we've decided that women aren't allowed!'"

Another synthesist from this period, Linda Fisher, a Moogist with David Borden's Mother Mallard's Portable Masterpiece Company, also recalls how gender issues got mixed up in the production and use of the new technology. Her comments add some perspective to Suzanne's experiences: "I mean, there were certainly men within those groups that had a very, what I'd say was traditional outlook [toward women], you know, who were just interested in how big their equipment was, or really weren't interested in the sounds that they were creating." Linda didn't see these attitudes as a barrier to taking on the synth as her instrument. She regarded it as a tool to be employed by anyone who had the capabilities. For her, a tool is neutral. Until altered by custom, tools do not come loaded with gendered implications: "It [the synthesizer] clearly came out of a male-dominated technology, in that sense. But it's a tool, like anything . . . There was a flexibility to it, I think, that would lend itself to anyone coming at it with any kind of approach."

As Linda developed her skill with the instrument, she discovered that the synthesizer was transformative and empowering. She visualized it as a countercultural force, and imagined her work as aimed at "subvert[ing] whatever the ruling, dominant outlook is." "For me, I don't know how other women feel about this, but having at my disposal the ability to make sounds that I've never heard before . . . that was great. And it was sometimes a little lonely being the only woman—get some other girls here!"

Linda found that the engineers and technicians were often more favorably disposed toward her than were her fellow male musicians. The engineers and technicians were less prone to show off, finding satisfaction in complex technical puzzles within the circuitry. The musicians, on the other hand, "really wanted to get out there and show what they could do with their big equipment." Linda found that male and female musicians displayed different attitudes toward the synthesizer: male musicians "would come to your concert and . . . they wanted you to blow their mind doing something new with technology." The women musicians tended to see the technology "as a leaping off point and not as an end in itself." These differences played out in the classroom as well. Linda taught the studio course in analog synthesis at Vassar College for two years and found that it was difficult getting women to sign up for it, and when they did, "they were usually very shy, didn't want to speak up, because they felt that they couldn't compete with these guys with racks of synths in their homes and, you know, [the men] knew all the terminology."

What Linda is pointing to here is the perhaps not so subtle gender dichotomies surrounding the synthesizer. By the time Linda was teaching at Vassar (the early eighties) the synthesizer had become a common sight in homes. The guys with their "racks of synths" would have an obvious advantage in terms of understanding the terminology and the technology. Having a synthesizer in your bedroom (along with a PC) was in a way an extension of the male hobbyist tradition of ham radios into a new era. The few women who were attracted to the synthesizer tended to be there for different reasons. "Those that persist, persist differently, and they persist for different reasons, because of really what the equipment can do for them, not just because it's a cool thing." The women's desire to explore the technology for what it "can do for them" is a persistent theme with all the women synthesists we talked with.

○

"Nobody Was Interested in What I Was Doing"

Perhaps the most innovative female synthesist of the sixties started her musical career as an accordion major at the University of Houston. Pauline Oliveros has played accordion since the age of nine, when her mom, a piano teacher, brought one home to increase her income. In 1953 Pauline got her first tape recorder and began taping found sounds, and at the end of the fifties she began making tape music after purchasing a SilverTone tape recorder from Sears and Roebuck. She was thrilled to discover variable speed recording by hand winding the tape.

Pauline composed her first tape piece at home, using all kinds of small objects that could vibrate: "I would record acoustic sounds using cardboard tubes as filters. I'd put a microphone at one end of a cardboard tube and a sound source at the other. I used different sized tubes to get different filter characteristics. Sometimes I'd clamp a sound source to the wall so the wall would act as a resonator and then record it at 3 1/2 or 7 1/2 inches per second and use the hand winding to vary the speed. I used a bathtub as a reverberation chamber."[2]

Pauline was a close associate of Ramon Sender and Morton Subotnick and played a role in the formation of the San Francisco Tape Music Center. Working in real time before it was thought feasible (1965–66), she produced sonic compositions by cobbling together the center's unused electronic equipment: sine tone and square wave generators connected to an organ keyboard, amplifiers, a mixer, a Hammond spring-type reverb, stereo tape recorders, a turntable with record, and two tape recorders in a delay setup.

The way she used the oscillators was particularly unusual: "I devised my own way of using these oscillators . . . I wanted a way to be able to perform, to work in real time with sound because I wasn't patient enough to make all those splices and wait to see if I got it right. So, I used tape delay. I set the oscillators at super audio, above hearing, and generated difference tones . . . heterodyning [putting the two frequencies together to produce a beat frequency] . . . The dials on those generators, they were very large, and the sweep it was very slow. But with difference tones you could make very minute changes of the dial and sweep the whole audio range."[3]

Pauline wasn't included in some of the decisions being made at the Tape

Center about which direction to follow in the new technology of sound. In terms of her approach to electronics, she told us, "Nobody was interested in what I was doing . . . in the technique." She was composing at the center when Don Buchla appeared with his prototype Buchla Box in late 1965. After listening to part of his demonstration, she went back to her studio to work on "Bye, Bye Butterfly" (1965):

> Now the male bonding in terms of technology continued and in the meantime I had devised my own way of playing the studio. I was quite happy with that, cause it was mine. Nobody else was doing what I was doing. So, they [Sender and Subotnick] mostly advised Don on that first synthesizer. He was the first engineer that came along who could execute what it is that they wanted . . . I didn't have a lot of interaction with Don, again, I mean it was co-opted into this male bonding thing. It wasn't that I was excluded purposely, but it was very hard for a woman to be a part of that discussion.

Pauline also composed on the original Buchla 100 at the Tape Music Center at Mills College, producing works with far-out sounds and names to match, like "Beautiful Soop" (1966) and "Alien Bog" (1967). Whether using her own equipment or the Buchla, Pauline's electronic music was nonmelodic and aleatory, making it challenging listening for the uninitiated. But her goals were personal. Pauline was less interested in working on "musical ideas" than working on her "mode of consciousness." Her music happened to be a result of this "mode." Pauline's intention was to have her music be unintentional, shifting its focus as soon as she noticed it becoming deliberate. It was part of her process for expanding individual consciousness through sound.

All three of these women synthesists had to overcome a variety of obstacles in following their love for a machine. The experimentation, creativity, and inventiveness embedded in the new technologies was exactly what they needed in their musical lives. The synthesizer's sounds provided an entrée to individual psyches, in order to disrupt the underlying fabric of societal expectations. The synthesizer when it first appeared was a heretical machine that allowed some synthesists to find a fulfilling musical outlet and to stretch cultural boundaries, but also to communicate the temper of the times through sound in the hopes of provoking critical change.

For Suzanne, Linda, and Pauline, music and political identities harmonized. When we asked Pauline how she and her music were affected by the countercultural scene in Haight-Ashbury, she asserted, "Well I don't think that it affected me—I think we affected it!" Linda explains further: "Living through that time. It was very intense politically, but there was also that sense that we could do anything we wanted. It was very idealistic, that we could be who we wanted, of course, always in totally politically responsible, totally creative [ways] . . . and there was this wonderful ideal sense that there was something better. You know, the anti-Vietnam protests, everything was this sense that your world doesn't have to be like this, there wasn't going to be war and people abusing one another with violence. There could be this good stuff."

The idealistic tenor of the times was part of their identity as musicians. They felt they were on the cutting edge, not only with sound but also with the impact those sounds were having.

◯

A Poetry of Sound

Suzanne's ultimate goal in working for Don was to acquire her own Buchla. She slowly built up her $8,500 synth, module by module, acquiring some of the basic ones while at Buchla's workshop. To have the system that she wanted, she realized she would need to earn more than the $3 per hour she was paid for stuffing Don's circuit boards. She first tried to become a recording engineer but found that "there was no receptivity at all for women engineers."[4] Finally she got a break from a "friend of a friend" who filmed commercials for Macy's in New York.

Macy's hired her to make sound signatures. She was able to continue living in California and ship her completed tapes out to New York (her producer was based in Milwaukee). She actually did her first marketable sounds using the Moog at Mills College because it was housed with recording equipment: "I took the scripts back with me and hid in the Mills College studio. Technically you aren't allowed to do commercial music there."[5]

The skill she was developing was in "sound design." "So it wasn't so much the note music as much as it was a poetry of sound—you know, what is the sound of a fur coat? What is the sound of a key chain? What is the sound of perfume? And developing metaphors in sound. The feeling, you know, the feeling you got listening to it. Was it soft and warm? Was it hard and cold? You know, so this poetry of sound is what I really brought to the industry."

With the money from these first commercials, Suzanne put together her Buchla 200. As she added modules, she found she was able to make more and more interesting sounds. Suzanne also found herself becoming closer and closer to the machine:

> Some people have a fear of technology, they look at this thing with all the knobs and holes and dials and things and go . . . "Oh, my God," you know. Whereas for me, it was like, "Okay, I'm going to get to know this. This is a living, breathing entity. It has desires and abilities, limitations and possibilities." And the process was getting to know the instrument. It was always in intimate and friendly rapport . . . And it was alive, you know, and you just have it on and you go and you interact and get to know it. You build up a relationship.

There is no doubt that the Buchla appealed to Suzanne's passionate sensibilities: "I always wanted it to be feminine and warm, and sensual. And the idea that this machinery could be sensual was a very feminine thing." The radical possibilities that Suzanne saw in the world of synthesis were

163

tied in with her countercultural values: "Well, it [synthesis] was a real counterculture as well because it was so new that it didn't have any precedents or limitations, so it fits right in the sixties." It was an exciting time when anything could happen, when people could and did fall in love with machines.

Suzanne, having by now completed her master's degree, frequently traveled to LA to look for work with movie producers. This was the same period when Paul Beaver and Bernie Krause were doing something very similar with their Moog. Eventually, Suzanne lived with her Buchla out at the beach in a guest house: "Everybody wanted that [unique sound], all the people who were writing film scores. I gave lessons to Leonard Rosenman, Dominic Frontiere, you know, big time Hollywood composers . . . But it was too complex." Suzanne realized that the complexities of the Buchla could be off-putting, and she did not carve out the same niche for herself with movie work as Beaver and Krause had done.

In 1974 she moved to New York, where she camped out with her Buchla in the Soho recording studio of Philip Glass. She thought that the sequencers on the Buchla would appeal to his style of music and offered to teach him: "[I taught Philip Glass] how to use it, and I tell you, he couldn't get it. He wanted to get it, and it wasn't for him." Glass himself has pointed to the limitations of the early synthesizer: "At that time synthesizers were not a practical performance vehicle . . . back then was before the era of polyphonic keyboards. We needed ten finger access and the only thing which offered that were simple electric organs by Farfisa or Yamaha."[6]

As a struggling artist trying to make it in the New York scene, Suzanne increasingly turned to sound-signature work for support. The commercial sound industry was based there and was a big-budget enterprise. Suzanne almost overnight discovered that she had become the Eric Siday of her generation. "I was immediately in the *New York Times*. New York loved me, I have to say, New York did love me." She became known for many industry trademarks: the GE dishwasher beep, the Columbia Pictures logo, the

Figure 24. Suzanne Ciani with her Buchla 200

ABC logo, the Merrill Lynch sound, the Energizer battery sound, the Coca Cola logo, and the Pepsi logo.

Suzanne by now was so enamored with her Buchla that in New York it was just about all she had for companionship. Her apartment contained no furniture, just her Buchla with its flashing lights sitting in the middle of the room. It was her partner, co-worker, and courtesan: "You know, a sound didn't just exist—everything was in flux. There was no 'is' there, it wasn't a static thing. Everything was shifting, everything was breathing. This instrument was, I mean, I had a problem, in a way. I remember when I went to New York and I was, I was scared, in a way, because I was in love with a machine. And I had this Buchla, and it was on, literally on for ten years."

Sometimes Suzanne would take the Buchla with her into a studio. The commercial producers were used to instruments that you pounded, plucked, or blew into, and operated with buttons, tabs, or keys. When someone wheeled in a synth with just knobs and patch cords, its oddity was disturbing; they didn't know how to relate to Suzanne and her Buchla. Yes, her reputation preceded her, she was the woman who could craft any conceivable sound, but this machine without traditional reference points was just too far out: "I'd walk into a studio without a keyboard and they'd go, you know, like they didn't know what to do, how to use it, what to write. Some of them just said, 'Do whatever you want . . . make the sound of a spaceship,' make a sound of whatever. But one producer said, 'Look, goddamnit! Get yourself an ARP String Ensemble [a keyboard synth made by ARP], that's what I want.'"

Still, Suzanne knew her instrument. She was the one who could say, "I made the sound of Coca Cola . . . The Coca Cola pop and pour was a logo sound for Coca Cola that was played all over the world and everybody knew [it]." She was in demand as the only synthesist on the East Coast who could create certain sounds. She used her highly developed sound skills and hard-earned intimacy with the Buchla to add the logos of Fanta and Sprite to her achievements, thereby turning her into the self-proclaimed "queen of soft drinks."[7]

Suzanne encountered the exact same problem that Beaver and Krause faced with their early use of the synthesizer. There was as yet no vocabulary to describe the sounds. Her patrons were trying to hear what they imagined would grab their customer's attention in the melee of the synthesizer's output. In a scene reminiscent of that described by Ray Manzarek when Paul Beaver first bought his synthesizer into a Doors session, Suzanne describes how the producer with whom she made the famous Coke ad worked:

> So you come up with a sound and if you touched one knob, suddenly everything was different. And these producers who didn't

know how to talk, nobody had the vocabulary for describing sound, he'd say, "No, no, go back, go back to where you were." So I'd move the knob back and he'd say, "No, no! It's not the same," because there were so many interactions—there were maybe fifty knobs contributing to one sound. The guy used to hit my hands— whenever he liked it I'd move it, he'd say, "Stop! Don't touch that, don't you touch another knob! Okay, record."

In working out a "poetry of sound," Suzanne and her customers were consciously searching for a particular sound signature that would elicit a special feeling in listeners. Although these were commercial ventures, the deliberate search for feelings packaged within sounds was at the cusp of innovation. Suzanne used her instrument "to create a poetry, a language, a musical equivalent of an idea, something that wasn't based in notes. You know, notes are little islands, and you can make a melody. Or you can have a chord. But now we had something else. You could make a gesture, a sweep." This way of making commercials was very different to the use of traditional instruments, "to start out you've got so many associations already with these instruments because we've had them for so long. And if I have a violin, you're going to hear a violin. Whereas, if I'm in a new domain, where you have no reference, it's so completely original."

It is perhaps not surprising that Suzanne is ambivalent about *S-OB*. Her attitude toward Carlos's work is shared by many synthesists who saw themselves as members of a radical musical movement. Suzanne respects Wendy's technical skills but doesn't feel that *S-OB* was the right vehicle to demonstrate what this new instrument was all about. Worse, *S-OB* misdirected the public's musical consciousness, constraining the way the instrument and its sound were allowed to develop. Although Suzanne had little interest in realizing Bach, she muses that, with a lot of exertion, she could have reproduced it on her keyboardless Buchla: "I could have done it on the Buchla—believe me, I thought of how I could do it—but it wasn't going to be easy." Indeed, the effect of *S-OB* was that Suzanne was inundated with offers. "After this happened everybody wanted, 'Okay, we'll give you a recording contract, but you have to do classical music on the synthesizer,' and I always refused."

Suzanne's ultimate goal was to make her own original compositions on the Buchla. Because she felt the synthesizer could be "feminine and warm," she believed she was using the Buchla in a unique way. On her first album, *Seven Waves* (1982), all the pieces are connected: "And the pieces are shaped also like waves, compositionally, and the idea is that this is a feminine form, the wave, you know, everything builds, builds, builds, and then releases. Okay, so it's a sexual form in a way, but it's a feminine architecture." Suzanne had completed two of the seven pieces and started looking for a recording engineer. Being in New York, she knew all the top studio engineers, but none of them could do what she wanted: "Men always had something to prove . . . the guys could tell you where to EQ the foot or the snare, or where to boost the mid-range for the trumpet—you know, they had all the answers. And here was something that had no precedent." What Suzanne wanted was someone who didn't have all the answers and was prepared to work with the material she had. She became almost desperate: "I had a vision, I thought I could make this music, and it's not there—and I cried, I was miserable." Finally, things changed when she met a woman engineer, Leslie Mona Mathis, whom she went on to work with for ten years, during which time Suzanne found new commercial success as a recording artist for her feminine style of New Age music.[8]

Suzanne is convinced by her experiences that women work in a different way from men in the new medium of electronics: "The woman comes to the work, the relationship is different. It's like, 'Okay, what is there?'—not what do I want to be there, or what should be there? . . . It's a little softer, open. And that's why I always felt that women were ideally suited to work in this technology, and I was very, very, very sad that that didn't happen . . . It's a man's world out there in the studio, a lot of it." Suzanne is here articulating what Linda had also observed: that women who work in this area come with fewer preconceptions about the technology. They are more prepared to listen to the material. One is reminded of Evelyn Fox Keller's analysis of Barbara McClintock, the famous geneticist. Keller noticed that McClintock too had a differently gendered way of working, a way "of listening to the organism" (in her case, corn) rather than imposing a view upon the organism of how it should behave.[9]

Unquestionably, the most remarkable thing about Suzanne is her special relationship with her synthesizer. But that relationship eventually came to an end. She discovered there was no resident electronic technician in all Manhattan, including the Audio Engineering Society, capable of fixing her Buchla. No one understood its insides. She had to ship it back to Don every time it needed repair, and this broke her heart: "You can imagine the psychological anguish that I suffered." Her synth's problems weren't helped by the fact that the Buchla almost always returned from its travels damaged again.

Suzanne eventually had to give up her Buchla. The emotional strain produced by her relationship with the instrument was overwhelming: "I was too emotionally attached, and, frankly, I was having a nervous breakdown, because when the thing was broken, I was broken. I was so attached to it that when it didn't work, I didn't work." So she started looking at other synthesizers, even though she felt intensely guilty about it: "It was like a lover, you know, being unfaithful. I went generic, I finally just went generic and I said, I miss all the magic and the uniqueness of the Buchla, but I can sleep at night now. I know that if this thing breaks, I can get another one, or someone who knows how to fix it, or they can send me a part." With the acquisition of a generic digital synthesizer, Suzanne finally found peace.

When Suzanne describes her Buchla as "living and breathing," is she merely making a category mistake or being overly sentimental? We think not. For Suzanne, the Buchla was not a machine in the sense of a fully accurate, fully controllable, mechanistic device. It was the very analog character of her synthesizer—its idiosyncracies and its imprecisions—that encouraged Suzanne to think of it in a different way. As she got to know the machine better, she formed a unique partnership with it, thereby carving out a new sort of analog human-machine identity for herself. It was an identity that crossed boundaries, that was hard to categorize, a perfect identity for a woman in a man's world who wanted to have it all.[10]

The paradox for Suzanne was that indeed she did start to have it all. The necessities of making a living, and ownership of the synthesizer itself (it was hardly cheap), demanded that she use her unrivaled skills by working at the sharp end of the very un-countercultural corporate universe. Like many people, Suzanne used the business world to satisfy monetary needs but always maintained her love for the "poetry of sound" and the radical machine that helped her produce it. The new way to sell and market via sound became a vast industry, with synthesized sound logos and commercials everywhere, and Suzanne's idealistic search for a poetry of sound was overwhelmed by the wider cultural forces she encountered. In a way, her uneasy relationship with the values of big business was shared by the iconoclastic machine she loved. The Buchla, too, never really found a cozy home in the corporate world.

9

Music of My Mind

The synthesizer [allows] me to do a lot of things I wanted to do for a
long time, but which were just not possible until it came along . . .
It's just that the ARP and the Moog give you another dimension.
They express what's inside your mind.

STEVIE WONDER

FROM TWO ENDS of the world Malcolm Cecil and Bob
Margouleff came together to build a hybrid synthesizer
such as the world had never seen (or heard). Beginning with a Moog III,
and adding on parts salvaged from a fire, second-hand electronic equipment, new modules from other companies like ARP and EMS, and modules they fabricated themselves, they created "the world's first, and largest,
multitimbral polyphonic analog synthesizer," better known as TONTO
(The Original Neo-Timbral Orchestra). Their enormous and evolving
synth was to be noticed by a very special musician, who would introduce
the sounds of synthesis to a new popular audience.

Born in 1940, Bob Margouleff lived in New York City during the late sixties and first heard the Moog in 1968 at the Electric Circus, an avant-garde
performance space in the East Village. He was finishing a film *Ciao!,
Manhattan* (1972), a documentary about Edie Sedgwick and Andy Warhol's
Factory, directed by John Palmer and co-produced by David Weisman, and
it was love at first sight: "All I know is the thing was like an epiphany. All I

knew was I had to have one." Soon after, he went to an AES convention and met Bob Moog. The synthesizer would be perfect for the movie's soundtrack.

Bob had been trained as a classical singer, but during this period he was jamming with Lothar and the Hand People: "It was the silver sixties in New York. And at the end of [*Ciao! Manhattan*] I just kind of ran out of money, and my parents really got radically pissed off with me, and disowned me. And all I ended up with was this big synthesizer, and me sitting on the street corner."

The silver sixties had the effect of emancipating Bob: "I flew in the face of every convention, and regaled in the results of it. I have to say, I did my share of psychedelic drugs . . . But it totally freed me and liberated me, so that when I became sober I was living in a new place." The "new place" for Bob was eventually to be Media Sound on 57th Street, where he moved himself and his Moog: "The studio was a ghost ship [always closed after six at night] . . . It was an old church, and the main room still had the old church organ in it. It was an electronic one, but a big one on wheels, on a console, and then my synthesizer on this big gurney I got from the hospital supply. And I would roll it out into the room there, and just really crank the thing up and really do the weirdest stuff I could possibly [imagine]."

○

"Oh, Good, He's Not Going to Be a Musician!"

Malcolm Cecil's trajectory into the world of music synthesis was from an unusual direction. He eventually became a respected bass player, but an early event first turned him toward a profession seemingly distant from music. He recalls this episode from the first day of school in Cricklewood, near London, when he spotted a piano. It was 1941 and Malcolm was four years old: "[I] start playing my little pieces . . . And everybody gathered round, and I was really embarrassed and said, 'Oh, well, you play.' And this kid says, 'I don't play.' 'You play.' 'I don't play.' So none of the kids play, so I said, 'Oh, the teacher will play for us.' And the teacher says, 'I don't play.' And all I could think about the rest of the day is this big revelation—everybody didn't play!"

Little Malcolm began to scheme, deciding when he got home that he now had all the evidence he needed to get himself out of his detested piano lessons: "And all I could say was, 'Everybody doesn't play. I'm not going to lessons anymore.' So my grandmother turned to my grandfather and said, 'Oh, good, he's not going to be a musician!' And my grandfather turned around to my grandmother and said, 'Of course not, he's going to be an engineer, aren't you, Malcolm?'"

As he internalized his grandfather's suggestion, Malcolm, like Bob Moog, realized that he had a knack for assembling mechanical gizmos, and throughout his childhood he continued to build electronic devices using parts he acquired from army surplus stores. This hands-on experience stood him in good stead; when he entered national service in 1958, he was assigned to radio training.

Malcolm kept up with his music (he came from a family of musicians), turning to the bass at age sixteen. He became a prominent musician as a resident bassist at Ronnie Scott's famous jazz club in London. We can see in his description of those times the characteristic musician's frenzy to earn a living by playing as much music as it's possible to squeeze into a twenty-four hour day: "We started to get individual American jazz musicians, famous ones, coming over . . . So one week I'd be playing with J. J. Johnson, the next I'd be playing with Stan Getz . . . until I started to get burned out on that, because it was five nights a week, and two of those nights, Friday and Saturday, were also all-nighters. So we did like the evening thing from eight to midnight, and then you came back at one o'clock in the morning and played until five in the morning."

Malcolm's health started to suffer, and he needed a steady job that didn't leave him with musician's hours. He found it at the BBC Radio Orchestra, which hired him as principal bass player. But his calling as an engineer would not go away. He is one of a very small breed of musicians who is just as comfortable with wiring schematics (circuit diagrams) and a soldering iron as he is with a jazz riff. He was always being asked to build equipment. He built one of the first four-track boards in England, designing it from a hospital bed for a friend who owned the Marquee Club and who wanted to record The Who live.

Malcolm's health, however, worsened; unable to play stand-up bass after an operation on his lungs, in 1967 he left Britain altogether and moved to South Africa as a concert promoter. His group included a black South African, an Indian vibes player, a white South African tenor sax player, and Malcolm on bass. Mixed-race concerts were against the law. He got away with it in Capetown, but when he reached Johannesburg he had a run-in with the police. Soon after, Malcolm was on his way to America.

◎

Moogists in Residence

Being unknown as a bass player in America, Malcolm found it easier to make his way as an engineer. He ended up with a six-week stint repairing a studio in New York. In this capacity he had a novel encounter with a fellow musician: "That was where I ran into Jimi Hendrix, he was working in Studio C, with this huge wall of Marshalls. And they called me up to the room because something was wrong with the machine. I came up and fixed it. And he's walking in and out of the door with this thing, creating this feedback sound, and that's how he was controlling the feedback—[he] was walking in and out of the door. I'm looking at this and going, 'Geez.'" After a successful repair job he eventually landed a job as chief engineer at Media Sound.

Analog synthesists all have stories about either the first time they heard S-OB or their first encounter with the synthesizer itself. Here is Malcolm's description of his first meeting with the Moog: "I walk into Media Sound and Studio A . . . I look up and see this big piece of equipment, weird. I look at it and it says, 'Moog' on it . . . Geez, this is the [instrument] that

George Harrison made that record [*Electronic Music*] on. I'm looking at it, and I saw it has filters, envelope generator—what the hell is all this stuff? So I go and I asked the people. 'Oh, this belongs to a guy called Bob Margouleff. Very weird guy, comes in at midnight, nobody likes him. You'll see, he's weird, he's very strange.' Can't wait to meet this guy."

Three or four days later Malcolm was staring at the Moog, and a voice came out of the darkness. Long-haired Bob Margouleff stepped into the light wearing "a fur coat down to the floor":

> BM: "Oh, you must be the new maintenance engineer. My name's Bob. I own that thing in there."
>
> MC: "What is it?"
>
> BM: "It's a Moog synthesizer."
>
> MC: "Oh really. It makes music?"
>
> BM: "Well, I don't know if it makes music or not. I'm not sure, but I'll play you some tape."

Bob, who "called himself the Moogist in Residence," stepped over to the Moog, took a key out of his pocket, and unlocked it:

> "I don't know how to operate this thing, but this is my tape." And he puts this tape on, this eight-track, and there's all sorts of stuff on it. Some weird stuff, stuff that's not that musical . . . And after playing this tape for about thirty minutes, he says, "What do you think?" I said, being the diplomat that I had been taught to be at BBC, I said, "Well, I think with a little judicious editing it might work." . . . He says, "Do you want to work on this for me?" I said, "I'll tell you what, you show me how to work that thing in there, and I'll show you how to work this board." "It's a deal," he says, shakes my hand, and that's how we became partners.

During the day Malcolm repaired and managed the studio; at night he taught Bob studio engineering and recording and Bob taught him synthesis: "And we slowly started doing some compositions together, and one thing led to another." They worked as a team, each taking turns while the other assisted. Bob's efforts might produce a sound and Malcolm would shout, "That's it! Let's get it on tape." And before the patch could disappear it was captured.

Bob recalls the music's transience: "The temporariness of it, the chaotic quality of it, the ability to create these most wonderful sounds that are there for a second and then go away, that you act on the thing in a very impulsive way, much as a jazz musician acts impulsively on his instrument. But the creation of the sound itself, the invention of the instrument itself comes very briefly to light out of chaos, and then it's gone again."

Sometimes they both played the synth at the same time, both contributing bass lines, melodies, or harmonies: "We would engineer the stuff together, you know, four hands at the console." They worked for themselves, not to make a record, not for fame, not even for money (at least, not during their nighttime sessions—during the day they found they could earn good money making sound signatures), but simply for the exhilaration of producing inimitable soundscapes. Early synthesists describe their intimate feelings for their equipment in provocative language; Bob will not refer to the Moog as a "machine" but refers to his synthesizer work erotically: "I would say it was right up there, right up there with [sex]. It was definitely boner time there, it was good. It was fun, it was very self-actualizing, very empowering."

The sounds they made were neither kitschy, funny, nor imitative. The soundscapes they built pushed the machine and their consciousness to the limits. As they migrated inward, the machine helped them move outward. This was a deliberate objective, and the way the politicized awareness of the sixties worked. It propelled you inwardly so you could be active outwardly. As Suzanne Ciani discovered, musicians came to the instrument and found a willing partner. Malcolm remembers: "What we tried to do is to make music that was intrinsic to the instrument . . . In other words, the instrument dictated a lot of how we went, rather than coming to it with pre-

Figure 25. Malcolm Cecil (right) with TONTO

conceived notions." The resulting music was an exchange of ideas between person and machine, both contributing to the final results. This may be why analog synthesists can readily recount feelings of love for their synthesizers.

At this point (in 1969) they were working on Bob's original Moog III and saw themselves as purists. Although they were tempted to add to the synth's sounds, there were no acoustic instruments on the final cuts of their first pieces. Their big break came when a friend of Malcolm's, Herbie Mann, the jazz flutist, visited the studio and Malcolm played him a tape of their compositions "Aurora" and "Cybernaut." Suddenly, Mann said the magic words, "Do you want a record contract?"

Now with a record contract, they needed a name. It led to the play on words by which they named themselves, Tonto's Expanding Head Band, although almost right from the start they affectionately referred to the Moog itself as the Lone Ranger's sidekick, TONTO. Tonto's Expanding Head Band is a pun, of course, but also contains within it an example of sixties sensibility animated by Marshall McLuhan's idea that technology would lead to a "global village." Malcolm: "Anyway, what the idea was, we were going to put down a track. Then we were going to send a tape to another synthesist and let them put something down, and then send it to the next one and let them put something down. And it was going to be, this was the Expanding Head Band . . . we never got to that. The tape never left our studio."[1]

The album, *Zero Time* (1971), complete with a psychedelic montage of planets, stars, and swirling nebulae on the cover, was groundbreaking. Its six cuts, including the warm fat bass sounds of "Cybernaut," the ambient washes of "Aurora," and the synthesized voices of "Riversong," made it an underground classic. Bob and Malcolm weren't aware that *Zero Time* had become a hit, "except somebody brings us *The Rolling Stone* and, lo and behold, there's a full-page article on how wonderful we are."

○

Conceived on a Tablecloth

As time passed and they needed additional modules, Malcolm would create them. Much of the design of TONTO was drawn over dinner on the paper tablecloths in a nearby Filipino kosher restaurant. Their synthesizer grew to mythic proportions—both in size and sound capability. The arched cabinets, designed by John Storyk so every control was within reach, meant TONTO was shaped so that our Moogists felt as though they were inside the machine.

We have seen other synthesists who, even if they didn't have the technical know-how to reconstruct the innards of their machines, recreated their synths by combining the modules that best fit their needs. In Jon Weiss's case, it meant linking both Moog and Buchla components in the same instrument. As Jon told us, "This kind of stuff, as far as I'm concerned, was in the true spirit of synthesizers, which is, you know, you take this equipment and you personalize it and you find things that you like and different ways of putting them together." For Don Preston, as early as 1965 it meant buying his own oscillators, combining them with a theremin, and constructing his own synthesizer, of sorts: "It was like a conglomeration of all kinds of electronic toys that you could buy, you know. And I put them all together and I think I had about forty oscillators."

As our Moogists in Residence got on with their compositions, Malcolm discovered aspects in the Moog's design that he wanted to improve: "As a bass player I wanted to bend pitch, I wanted to get some emotion into the thing, some loudness, some softness, and so on. We discovered that if we altered the voltage that went to the filter we could make the sound softer and louder, piano forte like, and also if we varied the voltage going to the oscillator a little bit we could get this pitch bending . . . I was hearing this stuff and couldn't get in between the cracks of the piano, so to speak."

To further "get in between the cracks," Malcolm went out and got himself a model airplane joystick. He "fixed it up" so that in one direction it controlled pitch, and in the other direction it controlled the filter. This controller finally gave Bob and Malcolm the feeling that they were able to create "musical lines." (This controller was similar to the one on EMS's VCS3.)

Malcolm found himself continually wearied by an instrument that would not remain in tune. It pushed him to make a device to correct the frustrating pitch drift of the Moog's oscillators:

> Now as you go higher and higher in pitch, depending upon where you set that variable resistor, you'll get more and more voltage. You feed it back, I bet you there's a way that you could compensate for that high frequency flatness, and sharpen it up. And it worked, and that's what's on the 920s [Moog's newly designed sta-

ble oscillators] . . . I do not want to make claims that I can't substantiate. But . . . I had it on TONTO . . . that was one of the first things that I discovered technically on the instrument.

They found that ARP would sell them individual modules, and this gave Malcolm the impetus to build his own modular accessories, for instance a voltage-controlled envelope generator (which Bob Moog didn't have) so they would be able to vary attack time with pitch. They soon got their hands on another Moog III at salvage prices whose case was burned in a Chicago dance hall fire, and they added two ARP 2600s, a ribbon controller, and two Moog drums. Eventually TONTO contained modules from Moog, EMS, Oberheim, Serge, and ARP, with Malcolm figuring out (pre-MIDI) how to get them all to "talk to each other."

Their Moog was now seriously gaining in dimensions, as Bob explains: "Our synthesizer was a whole bunch of little synthesizers. We had five or six filters, low pass and high pass. We had twelve, sixteen envelope generators. We had four keyboards. So we could play a lot of sounds at the same time." It was about nine feet long and they had to tow "the keyboard along on a little tea trolley to try to go from one end of the instrument to the other!"

Malcolm didn't see the keyboard as a constraint to musical innovation. He didn't think he had to play traditional twelve-note melodic music: "I particularly saw it as a freeing instrument, something [on which] we could be innovative rather than imitative." Malcolm wanted to leave behind polyphonic harmony and the twelve-note scale: "I was of the belief that this was the beginning of the music I'd been talking about all along, which had nothing to do with Western scales. In fact, 'River Song' is in seventeen tone . . . It was the first instrument I was able to tune to seventeen tone." Their goal was to make "timeless music"—"music that you couldn't put a period on, that could have been a thousand years ago, ten thousand years ago."

Occasionally they needed to imitate an acoustic sound—the story of

Figure 26. Inside TONTO: Malcolm Cecil (left), Bob Margouleff (right)

their attempt to imitate the sound of a bell, which became the gong on "River Song," is a classic look at the labors and triumphs of emulative synthesis. It also illustrates the thrill of crafting recognizable sounds, even to two synthesists whose primary love was producing the unfamiliar:

> We wanted this bell sound. And we figured out the envelope okay, that wasn't hard, you know, the strike and all that. But nothing sounded like a bell when we did it. So I said, "You know what, I've got this book, Helmholtz [*Sensations of Tones*], that I've been reading for years." I said, "I seem to remember . . . he analyzed the

sound of the big bell in Kiev, the harmonics, and he wrote them down" . . . So we dialed up the harmonics from the great bell of Kiev, exactly as Helmholtz had written . . . fed them into the mixer, put them through the filter, put the envelope on there that we'd already figured out, pressed the key, and out came this bell. I'm telling you, it happened. It was unbelievable! We were hugging each other, dancing around the studio. "We did it, we did it, we did it, we did it!"

The girth of TONTO presented great technical difficulties for live performance, but the lure of the stage grabbed our studio-bound experimentalists. The Moogists in Residence began offering an occasional live concert. During 1970–71 they performed a lunchtime concert at the Wall Street Church, with 300–400 people packed into the pews. It must have rocked the downtown financial district and been as close to a happening as the suited stockbrokers and money managers could have imagined.

Stage performance was right up Malcolm's alley, but Bob felt uncomfortable; and after their appearance on TV's *Midnight Special* with Billy Preston (1975), he told Malcolm he wouldn't do any more live shows. Besides, Malcolm's engineering sensitivities didn't combine well with Bob's nontechnical background: "[TONTO was] very difficult for me to use because Malcolm refused to put labels on anything. Like attack, you know, duration, T3, or any of that stuff. So [there were] these black panels with a thousand controls on the front of them."

○

Fulfillingness' First Finale

Zero Time fulfilled Malcolm's goal to produce music that was "timeless." The record also led a very special visitor to their door, dressed in a "pistachio-colored jump suit." Malcolm continues the story:

When I was working at Media Sound they had a third-floor apartment . . . that they provided me with . . . and [down in the street]

Ronnie Blanco [a bass player] . . . says, "Hey, Malcolm, can you open up the studio because somebody here wants to check out the synthesizer." "Okay, I'll be down." So I go down and get my keys out, open the door, and he brings this guy in, Stevie Wonder. He'd been working with Stevie. Played Stevie the album, told Stevie this is a keyboard instrument, you should be into this. Stevie had just turned twenty-one on May 13th [1971], about a week before.

Stevie first heard the Moog on *S-OB* and was immediately impressed.[2] Because of an exceedingly exploitive contract he had signed as a minor with Motown, giving him few rights to his own music, he had spent the previous few years composing many songs in his head and holding them there. Motown's rigid production structure, dogmatic musical values, and unwillingness to allow performers creative latitude, while leading to numerous recording successes, frustrated many of the gifted musicians who began their careers at the Detroit music factory. But at twenty-one Stevie was no longer tied to those agreements, and he was ready to let his songs out. His relationship with the Moogists got off to a great start: "We went up to Malcolm's apartment, and the Mellotron was up there, and we started improvising around on the Mellotron, and Malcolm picked up his bass and we were all laughing it up. We came down to the studio, and the next thing I remember it was four days later, and we had seventeen songs in the can."

Three or four of these songs came out on Stevie's first breakout album, *Music of My Mind* (March 1972). With Bob and Malcolm's production help, Stevie made three other hugely successful albums where the synthesizer wasn't used primarily for back-up sounds but became an integral part of the accompaniment: *Talking Book* (October 1972), *Innervisions* (1973), and *Fulfillingness' First Finale* (1974).[3] "Fulfillingness" was Stevie's name for Malcolm. By saying it was Malcolm's first finale, Stevie was telling Malcolm that the Moogist would be back to produce more recordings.

The three of them worked closely together: "I was programming, Bob was programming, I was engineering, Bob was engineering. We would switch hats at the drop of a hat. Whichever it was of us that had the idea—by then everything was flowing. It was just one flowing trip, with just the three of us in the studio, period." Bob and Malcolm set up the instruments so Stevie could easily reach them: "Piano, synths, drums, Rhodes, Clavinet, vocal mikes, etc. They were hot all the time. We had them in a big circle. Stevie would go from one to the other as needed."[4]

Malcolm describes how everything was "flowing" during the taping of "Boogie On, Reggae Woman": "One of us would work on the knobs, one of us, Stevie, would play the actual notes, and one of us would work on the keyboard. I would usually work either on the knobs or the keyboard things, switching in the portamento and switching out, watching his line, knowing what he was going to play, so the portamentos were in the right place, switching it in and out, turning the hold, no-hold on and off in the right places so the right effects were happening. So as a player, you couldn't have done it—one person could not have played that . . . It was the three of us together doing it that made the thing happen."

In the summer of 1972 Stevie became the opening act for the Rolling Stones on a major tour of the United States. He was performing with a mega-band and the public noticed. This had a tremendous impact on his record sales and the synthesized sounds that were sold with every album.[5] Black music and performers were hip.[6] It was a period when *Soul Train*, featuring black musicians, dancers, and sponsors, was ardently watched by the same kids that tuned into *American Bandstand*.[7] In another tour in 1974 Stevie appeared on the cover of *Newsweek*, and the magazine reported that "now the sheer creative power of Black music has pushed it into the mainstream."[8]

"Little Stevie Wonder: The 12 year Old Genius," as he had been called in Motown, had become a crossover artist.[9] His R&B and soul recordings, combining elements of gospel, rock 'n' roll, jazz, and African and Latin rhythms, allowed the synthesizer, thought by *Rolling Stone* magazine to be "the signature of his sound," to reach an unexpectedly large and varied audience.[10]

Figure 26. Stevie Wonder, Malcolm Cecil, and TONTO

Stevie signed a new contract with Motown in early 1972, making him the first Motown artist to win complete artistic control. Bob and Malcolm continued to work with Stevie; they moved TONTO out to Los Angeles and, with all the success they were having, they were able to employ a technician, improbably named Ulysses S. Grant, to work full time on TONTO. But according to Bob and Malcolm, their partnership with Stevie eventually became strained as music promoters and industry people were protective of their relationship with the mega-star. The Moogists found themselves with diminishing credits on the recordings. Malcolm: "We got a Grammy award in engineering, and two nominations. But our credits kept getting smaller and smaller, Stevie's credits kept getting bigger and bigger,

MUSIC OF MY MIND

and we were never taken care of from a royalty standpoint . . . we were called co-producers. And then it turned into associate producers, and then our names started getting smaller and smaller."

It turned out that *Fulfillingness' First Finale* was Fulfillingness' last act with Stevie. Since Stevie's blindness prevented him from following exactly what was going on in the studio, Bob and Malcolm are not convinced that he fully understood their skilled contributions to the production of his hits. The lines between engineering and musicianship were being developed during this period. Star power also has something to do with who gets the credit, as the program notes to *Music of My Mind* make clear: "This album is virtually the work of one man."[11]

Stevie's interaction with TONTO was both tactile and aural. Even Malcolm, who is confident about his auditory acuity, claims that Stevie was "the only person I ever met in my life who could hear stuff before me." It is likely that Stevie's fingers and ears could "see" TONTO as no one else could.

Bob and Malcolm today no longer work as synthesists. The machine that "once upon a time . . . represented the cutting edge of artificial intelligence in the world of music" today stands in Malcolm's studio-barn in Woodstock, New York.[12] Malcolm dreams of performing live with TONTO again. It introduced the Moog to an enormous new audience, and a generation of listeners found that they now had the sounds of the synthesizer on their minds.

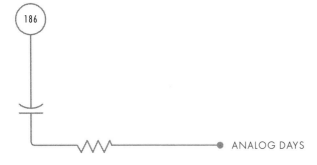

10

I play this thing on a wing and a prayer every night.

KEITH EMERSON

ᴮOB MOOG WAS LATE again. The four specially adapted synthesizers stood forlornly at one end of the sculpture garden, their electronic innards exposed as Bob and his engineers tended to them, dabbing solder here and there. People were already starting to file in. It was summer 1969, and New York City was about to experience its first ever live synthesizer concert.

When Bob had accepted the invitation from the New York Museum of Modern Art (MOMA) to conclude their summer "Jazz in the Garden" series, he knew he had to come up with something special. New York was his hometown, and the Museum of Modern Art was *the* place to showcase. It was here in October 1952 that Vladimir Ussachevsky and Otto Luening had given one of the first tape music concerts ever, including Ussachevsky's famous piece "Sonic Contours." Adding to the pressure, the Sunday *New York Times* had just carried a two-page feature on Bob and his synthesizer.[1]

What Bob came up with was a modification to the Moog to enable it to be played live. Turning the Moog from a studio oddity into a real instrument you could use in live performance had become a priority. Moog's sales reps, Walter Sear and Paul Beaver, were urging him to make a performance version they could sell. The demand was there: session musicians needed something portable to take from studio to studio. The record industry too wanted a live Moog; they had a new star instrument, but what good was it if you didn't have star performers to play it live?

Bob prized his ability to solve technical puzzles. The four specially designed Moogs on stage worked from a new pre-set box that allowed the musicians to change among six basic sounds at the push of a button. The engineers had made a rack of circuit boards, each of which contained numerous trimpots (very small potentiometers adjusted by a screwdriver) corresponding to many of the knobs on the synthesizer so you could set the parameters in advance. They had chosen the most useful patches and hardwired them into the Moogs.

○

The Moog MOMA Concert

S-OB had made the Moog famous, but now Bob and his synthesizer were facing the ultimate test: could it be used for live performance? No tapes, no endless adjustments, no excuses; just a few brave musicians, their Moogs, and an audience. And what an audience it was turning out to be. As Bob did the final checks, he watched amazed as the rows of chairs, set out for the few hundred people expected, filled up. Soon it was standing-room-only, and then climbing-room-only, as people clambered over the sculptures and into the trees to get a better view. Seated at the front were Bob's parents.

One year after S-OB, at the pinnacle of his fame, Bob was taking part in one of the most avant-garde events the museum had seen. Moog: "There were 4,000 people in that Garden . . . My parents . . . were very proud, it was their son up there, big event, famous . . . at some point a couple of people climbed up into the tree that was near where my mother was sitting. And after a while it looked to me like there was a little fire inside the tree, you could see smoke coming out of it . . . And after about a half an hour or so my mother says, 'Robert, what is that wonderful smell?'" That "won-

188

Figure 28. Live at MOMA, 1969

derful smell" drifted over many sixties events. At MOMA that night, two weeks after Woodstock, nearly everyone was high whether they smoked dope or not.

Opening the concert was Herb Deutsch and his quartet, featuring jazz pianist Hank Jones. They performed electronic be-bop jazz. The highlight of the evening was a free-form jazz group led by Trumansburg musician Chris Swansen, which included John McLaughlin (of later Mahavishnu Orchestra fame), pianist Hal Galper, and drummer Bobby Moses.

Jon Weiss was on the mixing console that night: "It was totally wild . . . the guys hadn't had much time to work on the machines, the musicians didn't know what they were doing, so they were just kind of winging it. And something happened, and there was music, it was cranking, it was cranking, and then all of a sudden the power went out. But nobody cared . . . and everyone was cheering—it was like a happening, that's what it was."

189

LIVE!

Bob Moog, seated near the front, was in a better position to see what *had* happened: "They got to wailing, just making huge waves of sound, and it got raunchier and louder and more dissonant. It just built up and built up and built up, and the crowd was going nuts, and somebody who was standing next to this electrical outlet decided to stand on the box that the outlet was in, so she could see better. She slipped off the box, knocked the power cord out of the socket, and everything went dead very abruptly, and that was the end of the concert."

By the time the music stopped, no one cared. The crowd loved it. The concert was less of a success in reviews. The *New York Times* panned it, but this is to miss the point.[2] It was the very presence of the Moogs that made the event so special. Like watching the first airplane get off the ground, or the first house light up with electricity, people had come to wonder as much as to listen.[3] And the new wonder was the Moog.

○

Mother Mallard

Other enterprising musicians were soon venturing into live performances with the Moog. One of the first ever live synthesizer groups, Mother Mallard's Portable Masterpiece Company (later Mother Mallard), was formed in late 1968 by David Borden.[4] Borden had been given free use of Moog's Trumansburg studio and had started to compose for the new instrument. As a dance composer, he was much influenced by Merce Cunningham and the musicians around him, including John Cage, Gordon Mumma, David Behrman, and David Tudor. He became convinced that electronic music was meant to be performed live. The type of music he made later, with its exploration of soundscapes and repetitive patterns in subtly varying timbres, became known as "electronic minimalism." According to Borden, "It was before minimalism was invented as a word, and so then people would lump us together with that." Although Borden is perhaps the missing link between minimalism and electronic music, he is much less well

Figure 29. David Borden with Mother Mallard, 1973

known than other minimalist pioneers like Steve Reich, Philip Glass, and Terry Riley, who embraced synthesizers later.

When he and Steve Drews (and later Linda Fisher) first started performing live as Mother Mallard, it was a humbling experience. They found that the Moog itself was the main attraction. "So we'd go to a place and they'd advertised it not as me, or the band, it was just 'Come see the Moogs.'" Mother Mallard's first concert took place in May 1969 in Barnes Hall, Cornell University. Each of the three performers had multiple keyboards. They

LIVE!

Figure 30. Mother Mallard, 1975: Judy Borsher (left), Steve Drews (center), David Borden (right)

used modular Moogs and later, in 1970, the very first prototype of the Minimoog, which sat on top of an electric piano, the latter being needed to keep the synths in tune. The first Minimoog ever to be used in a concert was that prototype (for a piece called "Easter" premiered on Easter Sunday 1970, again at Cornell). Some of Mother Mallard's material was improvised, some was scored, and sometimes they recorded a tape of themselves live which they would then improvise over later in the same performance.

The new technology required performing musicians to evolve completely new sets of practices. Mother Mallard typically used standard musical notation for their compositions and made special shorthand patch diagrams to delineate the labyrinth patch changes—sometimes as many as a hundred during a single piece. Other musicians adopted other solutions. For instance, Don Preston of the Mothers of Invention told us he never wrote down patches. "And the reason for that was I felt it was more important to know the instrument well enough so that I could go back and recreate that sound. It might not be exactly the same, but it would be close enough . . . in doing that I would become more familiar with the instrument." Bob Margouleff and Malcolm Cecil exploited TONTO's vast size and complexity to keep the show going. One of them was able to continue performing on one section of the instrument while the other switched the patch. If TONTO went belly-up (as it did during their Wall Street concert), they simply switched to playing a tape of their compositions—no one in the audience could tell the difference.

Mother Mallard practiced their patch changes in rehearsal: "We'd go into army drills." This produced a remarkable scene: "We used to have rehearsals where we didn't play any music, we were just practicing the patching for the pieces . . . So we got it down to five minutes, five-to-seven minutes between pieces." During the five-to-seven minute interval where Mother Mallard would go through their silent patch change choreography, they would show classic cartoons from the thirties and forties. David recalls that, for some, the cartoons may have been the best part of the evening. "And I heard someone go out once and say, 'You know, it was worth it just coming to see the cartoons.'"

○

On a Wing and a Prayer

Bob Moog had not designed his modular Moog synthesizer with live performance in mind. As a result, synthesists could never get up on stage and hope that everything would work properly. They had worries beyond the broken strings and dead mikes of conventional musicians. One of the biggest problems lay right at the heart of the synthesizer—the oscillators. These refused to stay in tune and were particularly sensitive to temperature. This was the single biggest headache that live use of the Moog presented.

Keith Emerson: "I had my faithful roady Rocky tune the instrument to A

440 just prior to the audience coming in, but once the audience came into the auditorium and the temperature rose up then everything went out of tune . . . I was playing away with my right hand and tuning with my left and I finally got around to using a frequency counter so I could switch off the audio and be able to look at the tuning before I did a solo."

When things went totally wrong, Keith went back to his sixties roots: he had a freak out. "You could get away with it by just sort of really screwing the instrument just totally up and just freak out with it. Pitch bending and wailing, and the rest of the band would sort of like just have to keep playing through it, in the hope that they recognized some sort of signal to come back in and play the rest of the arrangement." Sometimes a Cagean moment occurred by accident. Keith: "It would pick up radio stations, taxis driving by; on one occasion it was so bad what we did was we covered the instrument with silver foil so it looked like we had a huge Christmas cake on stage."

Small amounts of dirt and moisture were the circuits' enemies. Borden learned early on from the engineers that most problems were "something dumb" and not some intricate circuit problem. One of Mother Mallard's later players, Chip Smith (who had played piano with Chuck Berry) invented the "FAWTS method" of fixing problems in live performance. The FAWTS method is "Fuck Around With The Screwdriver." Bill Hemsath (a Moog engineer) had shown them this trick to get the contacts on the Minimoog circuit cards to tighten up. Borden: "So when there were a lot of people around he'd say, 'Let's use the FAWTS method.' I said, 'Oh yes, the FAWTS method.'"

Outdoor concerts were a particular problem. With the whole "back to nature" ethos of the sixties, there were many such venues. Borden: "We played at some of the communes around Ithaca . . . [the engineers] figured out a way to shield it somewhat, with I think just glue . . . but still if the sun was on it [it wouldn't stay in tune] . . . So we learned how to play and tune

194

at the same time." Although everyone likes to tell tales of the equipment going wrong—it makes for better stories—in defense of Bob Moog and his modular equipment, it should be pointed out that on one famous occasion a sudden squall blew over Keith's modular Moog and it filled up with water; but two days later, after being dried out, it still worked.

As with the studio use of the Moog, the question arose as to what genre of music the live Moog would be best suited for. And here to some extent the debate between exploring new sounds or emulation repeated itself. The Trumansburg factory had in effect two house bands, Mother Mallard and the Chris Swansen Trio, and each took a very different approach toward the Moog. Swansen used the synthesizer to make, as Bob described it, "switched-on jazz." Borden, firmly in the camp of the experimentalists, has his own impressions of the difference between the bands: "I was merely interested in letting the Moog be Moog . . . The Swansen band . . . [played] some kind of fusion stuff, and it was emulating other instruments, which it was quite good at . . . and he [Swansen] thought he was going to be the Walter Carlos of jazz."

Borden's lack of public recognition is somewhat of a puzzle. He has steadfastly refused to move to an urban center, finding the rural setting of Ithaca to be an inspiration for his music. In early 1970 Borden nearly got the big break. The film producer Billy Friedkin had heard Mother Mallard on the radio in NYC and thought his music would "fit well" with his new movie:

> He . . . hired us to do the music for the *Exorcist*, more or less, you know, verbally. And then we went to the cast party at the end . . . saw the mechanical bed, and then I get the call three weeks later, he's saying the film is much too melodramatic . . . he's gonna have to revert to what he usually uses, which was a montage of music from various sources, but if I could think of anything, send him

195

something. So I sent him three things and he used them all in the film . . . When the exorcist first arrives on the scene I put the ominous drone down.

After the success of *The Exorcist* (1973), Friedkin invited Borden to Los Angeles to make movie scores, but Borden preferred to remain in Ithaca. He was also never interested enough in the pop world to turn Mother Mallard into something like a Tangerine Dream or a Kraftwerk.[5]

Although people are slowly realizing the significance of Mother Mallard's early music, such as the mesmerizing "Ceres Motion" (1973) with its heavenly sound and gently rolling synth drones, and "Easter" (1970) with its exploration of effervescent staccato timbres and disquieting sound collages, the early categorization of them as a novelty Moog band was sometimes hard to escape. Critics derided electronic minimalism as boring because "nothing happened." Mother Mallard finally stopped touring in the 1980s after Borden heard someone in the audience describe them as "just another synthesizer band." Today they still perform in public and in 2001 reformed as an all-synthesizer trio.

◯

The First Moog Quartet

Another way of using the Moog live was pursued by Gershon Kingsley, a German-trained classical musician. With "French tape wizard" Jean-Jacques Perrey, he had collaborated on a plethora of pop tunes. Their album *The In Sound from Way Out* (1966) was spliced together from innumerable taped sounds from acoustic instruments and a Jenny Ondioline.[6] It was one of the first popular recordings to draw attention to electronic music and became somewhat of a cult album. It led to much work for the pair making radio and TV jingles, and their "Baroque Hoedown" was taken up by the Disney theme park for the nightly Main Street Electrical Parade. Their second album, *Kaleidoscopic Vibrations: Spotlight on the Moog,*

Figure 31. Gershon Kingsley, 1969

197

LIVE!

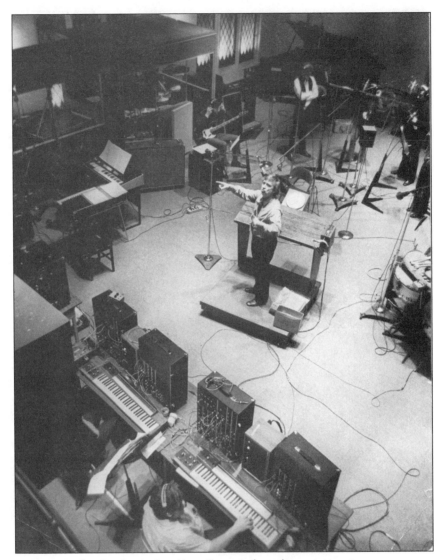

Figure 32. Gershon Kingsley conducts Moog Quartet in the studio

composed on a smaller modular system, was one of the first pre–S-OB recordings to use a Moog.

Kingsley saw the live potential of the Moog after attending the MOMA concert: "I saw these four Moogs, just demonstrating sounds, not actually music." He decided to form his own First Moog Quartet. Through his connections with the famous New York impresario, Sol Hurok, he ended up with a booking at Carnegie Hall for January 1970. There was, however, a problem: "no synthesizers, no players, no program, nothing." To buy the synthesizers (Moog Model 10s) he did a deal with a record company whereby he got an advance of $30,000 and they were granted exclusive rights to record the first ever synthesizer concert at Carnegie Hall. To get players, he put up ads at Julliard and other music schools. After auditioning 150 candidates, he ended up with a core of five musicians: "Nobody knew how to play the Moogs so I had to teach them the basic things, you know. It was unbelievable . . . this concert was in January."

The First Moog Quartet performed at Carnegie Hall (accompanied by movies and dancers) to less than rave reviews, receiving an "outright slam by the *New York Times*."[7] Gershon maintains a dignified memory of the reviews: "The reviews were mixed . . . the *Times* hated it, the *Post* loved it. It was always controversial." One lasting impact of that concert was that David Van Koevering, in attendance with Bob Moog, became convinced of the live potential of the Moog and embarked upon his own project to bring the Moog to the people. The First Moog Quartet toured but was never a favorite of the critics, who felt the music was always either too kitschy and gimmicky or a pale imitation of the conventional genres it tried to emulate.

One unexpected spin-off, however, was that it led to "Popcorn," one of the best-known Moog pop tunes ever. It also made Gershon a mint of money. He claims he wrote the song, with its catchy melody based upon a Jewish folk tune, in about two minutes. It became a staple of the First Moog Quartet repertoire and would probably have been forgotten if a member of the group, Stan Free, had not gone on to record it with another group, Hot Butter. Hot Butter's bubblegum version became an international hit, selling over a million copies in Germany alone. "Popcorn" can still be heard today in advertisements (for example, a recent one for Nike), it has over five hundred cover versions, and it is one of the most immediately recognizable of all Moog tunes.

◎

Keith Emerson

It fell to a rock 'n' roller who prides himself on his lack of knowledge about technology, Keith Emerson, to discover how to make the Moog into an exciting performance instrument and one that people would flock to see even after the novelty had worn off. Keith was the first major rock musician to take the Moog on the road. His "Monster Moog" became his trademark instrument and helped turn him and his group, Emerson, Lake and Palmer, into international super stars. ELP were at the vanguard of a new British invasion in "progressive rock," and in 1971 and 1972 topped the *Melody Maker* poll for best British group, ahead of the Beatles, the Rolling Stones, and The Who. The glam and pomp rock excesses of ELP may have obscured Keith's achievements. He was not only a keyboard virtuoso but also an innovator. Keith Emerson, Rick Wakeman, and the like did for the keyboard what Jimi Hendrix did for the guitar. They turned it from a background piece of furniture into an instrument where the rock keyboardist could become a soloist and center of attention on a par with the guitarist.

Keith was born in 1944, and his exposure to music began early—both his parents were amateur musicians. His father played accordion in a dance band and insisted that Keith (unlike himself) should be able to read music and be versatile enough as a performer to fall back on music as a sideline career. He learned piano from the age of eight, mainly taught by a succession of local teachers in Worthing on the South Coast of England where the Emerson family lived. His talent blossomed, and he soon entered competitions. At age fourteen he even turned down an offer of a place at London's Royal Academy of Music.

One early influence was skiffle. A product of Britain's postwar austerity, skiffle was a kind of do-it-yourself music using a tea-chest, broom stick, and strings for bass, a washboard for rhythm, and a comb and paper for melody. Postwar Britain, reeking in poverty and dying of boredom, was the teenage breeding ground for a whole generation of innovative rock musicians, including the Beatles, the Stones, Cream, and The Who.

Most rock stars of Keith's generation were influenced by the blues. Keith, though, had an unusual American influence, John Cage. "I was aware of John Cage, who . . . was sticking things inside a piano like ping pong balls and various other things, so I thought this is a light relief from what I was having to learn. I experimented with it sometimes, until I was told off by the [school] music teacher—you know, 'You're not allowed to do that!' But I was persistent, I thought, well if you're not allowed to do something, I'm going to damn well go ahead and do it!" The role of chance and experiment in Cage's music was to be a lasting influence on Keith's career as a musician.

○

Organ Abuse

After school Keith lived in Brighton, took a job as a bank teller, and worked in various bands. At this stage of his career he saw himself as a jazz pianist. He remembers Floyd Cramer's record *On the Rebound* as being a big influence. Playing in sleazy Brighton dives for drunken sailors, he enjoyed improvising around the themes of current pop hits.[8] He bought his first Hammond organ (an L-100). He wanted to play like Jimmy Smith and sound like Brother Jack McDuff but was disappointed when "I got it home and it didn't quite sound like Brother Jack McDuff's."

In those days Keith was a typical jobbing musician who could turn his hand to any genre of music; one moment he would be playing organ for a bingo session and the next a dinner dance or a club or jazz date. Eventually he joined an R&B band, the T-Bones, who had a residency at the Marquee Club in London. It was at the Marquee that a little-known organ player, Don Shin, took the back off the Hammond and "got out a screwdriver and started making adjustments while he was playing."[9] Keith himself started to get inside his Hammond organ. He did everything from "tapping the valves of the organ to playing the reverberation springs," and made "air raid sirens to machine-gun sounds, by sort of rattling the reverb chamber."

Keith discovered the effect on his audience of these sounds by chance. He had left the T-Bones to join the more purist blues band, the VIPs (who later became Spooky Tooth) and was touring northern France with them when a fight broke out in the dance hall. "The guys said, 'Keep playing.' So I sorta joined in on the Hammond." He made some of his favorite explosion and machine gun sounds, "going completely crazy and the fight stopped in the audience and they all looked up [acts out expression of amazement] . . . I played another concert and the guys in the band said, 'You'd better do that again! Because that went down well.' So I did go on and do that again, It was great." Keith was now on the way to developing one of the most flamboyant stage acts in rock. The idea of the "keyboard hero" was born.[10]

Like most aspiring rock musicians at the time, Keith could hardly escape the sixties. He was right in the center of it in London: "All musicians were being very experimental in those days. Expressions such as 'freak out,' underground music you know, that was part of the genre . . . I think we were all looking to play outside our instruments . . . to explore the dimensions of what could be achieved." The Nice, Keith's band after the VIPs, was formed by ex-T-Bone player Lee Jackson, to support R&B recording artist Pat Arnold. In September 1967 The Nice separated from Arnold and took part in what was billed as the "first psychedelic tour of the UK," performing with the Jimi Hendrix Experience, the Move, Pink Floyd, and Amen Corner. Hendrix, with his innovative style of playing and use of feedback, was another influence on Keith. At one point, Hendrix almost joined ELP.

○

An Encounter with a Baroque Gentleman

One of The Nice's big hits was their version of the *Third Brandenburg Concerto*.[11] It went to number 10 in the British charts, which was remarkable for a classical piece and evidence of the same crossover between classical and rock that boosted sales of *Switched-On Bach*. Keith was thus all ears when he walked into a Soho record store one afternoon and the owner said, "Well you've done the Brandenburg 3. Check this out," and preceded to play him *Switched-On Bach*. "My God that's incredible, what is that played on?" The owner then showed him the album cover. "And there was a gentleman dressed up in a baroque outfit and a thing that looked like a telephone switch board. So I said, 'What is that?' And he said, 'That's the Moog synthesizer.' My first impression was that it looked a bit like electronic skiffle."

The Nice's manager, Tony Stratton-Smith, did some research around London looking for a Moog and soon tracked down Mike Vickers of the Manfred Mann group. Keith had a go at playing Mike's Moog and soon found he couldn't. "'No, no, no! You can't play chords on it, you can only play one note at a time.' Okay, so I got into it, it was great." Keith was particularly taken with the portamento control because that was the effect he had been trying to achieve on the Hammond organ by switching the tone generator on and off. "This thing went '*wooeeeeee*' and you could control it."

The Nice had a sold-out concert to play with the Royal Philharmonic Orchestra at the Festival Hall, London. Keith arranged to borrow Mike Vickers' Moog, and Mike ended up doing the patching, as in "the kabuki theatre, like the shadows pop up from time to time and change, so Mike would jump up from time to time and change the patch cords."

The night of the concert arrived, and Keith showed up bedecked in a specially made silver lurex spacesuit. After running through their "Five Bridges Suite," Keith took the controls of the Moog to play Richard Strauss's "Thus Spoke Zarathustra"—the theme from the Stanley Kubrick film 2001: A Space Odyssey, which had just been released. The piece was a show stopper. "We did an excerpt of that using a lot of legato . . . and the Moog fitted in perfectly. We had the white noise rushing from left to right across the stage . . . And it was amazing to look at the audience's face; it was complete bemusement . . . some people didn't think that that sound was coming from the instrument at all, a lot thought there was a tape recorder back stage . . . I thought this was great, I've got to have one of these."

◎

How Not to Be Unfair to the Beatles and the Stones

Keith even tried getting one for free, or at least his manager, Tony Stratton-Smith, did, by writing to Walter Sear. Sear was unimpressed, writing back that "we have never offered instruments to groups for promotional use, first because of the cost of the unit and secondly, because of the small size of our company. It would also be quite unfair to the groups (such as the Beatles, Stones, etc.) in England who have purchased the equipment."[12] Sear also issued a stern warning about "the time and training" necessary before the Moog could be used to make any sounds. Was a mere pop artist really worthy of such an instrument? This was Walter's message. As Keith notes, "It's hardly a promotional letter, it's not sort of endorsing the instrument."

Jim Scott, who had the task of getting Keith's instrument ready for delivery, recalls, "Well, Al Padorr, the marketing manager, sent around this little blurb about this musician nobody had ever heard of named Keith Emerson . . . and it was going to be delivered to England. And of course, to England, how's this guy going to know how to use this thing? So I drew up a diagram of a standard patch that was to be used—put the patch cords there and don't touch them." Back in England, where a few people had heard of him, Keith was about to embark upon the project that was to change his life and

204

propel him from cult band to international rock star. Dissatisfied with playing in The Nice, Keith teamed up with Greg Lake, the King Crimson bassist and vocalist, who was also dissatisfied with his band. They were joined by former Atomic Rooster drummer Carl Palmer, to form Emerson, Lake and Palmer. They soon had a record deal from Atlantic and a budget to buy instruments. Now Keith could actually pay for his Moog.

The big day arrived. Keith: "It cost a lot of money and it arrived and I excitedly got it out of the box stuck it on my table and thought, 'Wow That's Great! a Moog synthesizer [pause] How do you switch it on?' . . . There were all these leads and stuff, there was no instruction manual." Keith's synthesizer was one of the four that had been used at the 1969 MOMA "Jazz in the Garden" concert. It had the pre-set box so that about six sounds could be changed at the push of a button. It was unusually complicated to set up because of this additional module.

Mike Vickers once again came to the rescue, and Keith left his synth in Mike's capable hands. "Mike got back to me a week later and he said, 'Yeah, it was tricky, but I've got six sounds.' And those six sounds became the basis of the ELP sound actually, I mean, I worked off of that . . . particularly in the recording of *Pictures at an Exhibition* (1972) and 'Lucky Man.'"[13]

○

Hoedown

Keith did go on to find some new sounds himself. "I just experimented by taking a patch cord and plugging it in somewhere else and seeing what happened. I had a pretty good knowledge of the instrument by that time, and I knew where the sound source went, where it should be directed to and a lot of the dials and things." Sometimes he found sounds by accident, such as the famous sound at the start of "Hoedown." "We'd started working on that arrangement and then I hit, I don't know what, I switched a blue button and I put a patch cord in there, but anyway *'whoooeee.'*"

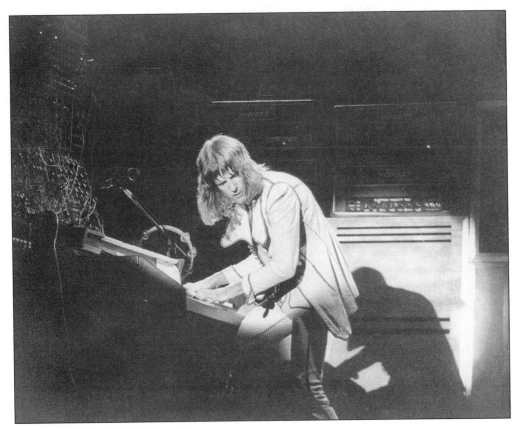

Figure 33. Keith Emerson with Monster Moog

Keith composed on the piano and used the Moog for overdubbing to "add a solo or a little bit of color here." As a keyboard player who had trained on piano and moved on to Hammond organ, Keith had to overcome the Moog's monophonic keyboard. He dealt with this by sometimes using his left hand for a solo and then filling out on another instrument, such as his Hammond organ, with his right hand. He specialized at introducing classical techniques like intricate counter-rhythms and contrapuntal lines. He had already developed novel techniques for playing two Hammond organs at the same time, one with each hand. "I kind of approached it the same way as I think a trumpeter or sax player would go about soloing. And having a good left hand I was sort of able to accompany those solo lines, and that kind of filled it out." Thinking of the Moog as a saxophone or trumpet was the key to his success. With portamento and pitch bending, not to mention the ribbon controller, he could make the solo stand out and come alive. Keith found that the monophonic keyboard, so often seen as a shortcoming, if used in the right way, could produce dramatic effects. Indeed, when polyphonic synthesizers came along in the late seventies and eighties, the keyboard in rock once more started to revert to the background, to be used for fills and atmosphere rather than for soloing.

Keith's explorations on the Moog were part of a keyboard renaissance in rock. Other rock groups like Deep Purple (with Jon Lord on keyboards) used the electric organ as part of a new "heavy" rock sound. The Mellotron, an analog precursor to digital samplers, with lush string sounds and choral effects, also had an enormous impact; it was used by the Beatles on "Strawberry Fields Forever" and taken up by the Moody Blues, Yes, and Genesis. King Crimson used a Mellotron Mk2 to produce their stunning debut album, *Court of the Crimson King* (1969). Rick Wakeman, former prodigy at the Royal College of Music (dismissed for spending too much time playing rock sessions), was perhaps the most versatile of all the new rock keyboard players. He was a member of the British progressive rock group Yes, before launching a successful career as a solo keyboardist. He, like Keith Emerson, achieved the distinction of having his own sound inscribed on synthesizer sound charts (such as those offered with the Minimoog), enabling other keyboardists to emulate his sound if not his virtuoso playing.

○

Lucky Man

Keith to this day is somewhat embarrassed about the hit that made ELP and the Moog famous in rock circles. "Lucky Man" was the first recording on which he used the Moog synthesizer. As he told us, "That's a solo that

I've had to live with!" "Lucky Man" was written by Greg Lake as a fill for the first ELP album. It is an acoustic ballad, a sort of ironic antiwar protest song about a "lucky man" who has everything, but who goes "to fight wars" and who is shot and is "laid down and he died." The chorus comes in with "Oh, What a Lucky Man he Wa-as." Greg Lake, who also produced ELP, had overdubbed lots of acoustic guitars and vocals and suggested to Keith that he do a solo. Keith found there was a space at the end of the tape where it kept repeating:

> So OK I set the Moog up in the back, 3 oscillators all with sine waves, and using my favorite portamento, patching arrangement . . . They ran it through the tape, and I was just really just jamming around . . . I got the thumbs up from the control room. Great! The look of excitement on their faces. "What? What do you mean?" "That's it man, that's the one!" I said, "No, no, no, that's dreadful," I said, "Let me have another go." And Greg was saying, "No man come in and have a listen it's unbelievable." Alright, "I'll come and have a listen." So I went and had a listen to it and I said, "Can't I just do one more? Is there an additional track?" "No man, all the tracks have been used up, we can't use any more. That is the solo!" So I was devastated.

Keith may have been devastated, but he had just laid down one of the most influential synthesizer solos in rock 'n' roll. The power of the solo comes from the contrast with the ballad format of acoustic guitars and vocals; the fat bass sound slowly comes in and off goes Keith on a yowling solo. The sound of Keith's three slightly out of tune oscillators was to become one of the definitive Moog sounds.[14] The meandering solo with lots of glissando does sound a little like someone who has just discovered the instrument, but its very "over the topness" is exactly what gives it its charm and, of course, was de rigueur for guitar solos at the time. Today, it almost sounds like an ironic take on what a cheesy synthesizer solo is meant to be about. It ends by panning from one speaker to the other—"Lucky Man" became popular with hi-fi storeowners as a way of demonstrating the power of their latest stereo equipment.

Played through stereo headphones, or a good speaker system, or through the huge PA system that ELP used in concert, the effect of the Moog with its enormous sonic energy is powerful. Another place where synthesizer recordings were particularly effective was in the car. The growth of FM radio in the sixties meant that songs like "Lucky Man" got lots of air-play. The solo sound of the synthesizer stands up well in the acoustically challenging environment of the modern automobile. The synth riffs on tunes like "Lucky Man" or the introduction to "Baba O'Riley" by The Who or even Gershon Kingsley's "Popcorn" sound convincing when played over a poor speaker system in a noisy environment—the sound cuts through dramatically. Any lack of subtlety does not matter so much in these environments.

It was hearing "Lucky Man" on a car radio as ELP drove from JFK airport to their hotel at the start of their 1970 U.S. tour that made Keith realize what he had done. "We had the radio on and they were playing the single 'Lucky Man' to my horror. I know we were called Emerson, Lake and Palmer but we were hardly Crosby, Stills and Nash and I couldn't sing."

◎

Bob and Keith

Bob Moog and Keith Emerson have formed a special relationship over the years. Bob, a shy engineer with a PhD and pen protector who was raised mainly on classical music, seems an unlikely person to get together with a fully-fledged knife-throwing, armadillo-dressed rocker from the London psychedelic scene. But appearances can be deceptive: it turns out that Bob likes to perform, and Keith—in private life—is shy and modest and has a very wide taste in music.

Bob had found his performer, and Keith had found his instrument. And young men everywhere had found something they could go "ape

LIVE!

Figure 34. Bob Moog and Keith Emerson

shit" over—a keyboard hero with a monster, gleaming piece of technology. (Keith told us that sometimes he would add to the myth of the "Monster Moog"—17 square feet and 550 lbs—with some blank modules and an oscilloscope purely for visual effect.)

Both men seem to have intuited early on the importance of linking their careers. When ELP's first album came out, Keith remembers Bob coming over to London to do an interview with him for *Melody Maker.* Bob also remembers that occasion; the gift of a test pressing of that album, on which the track "Lucky Man" appears, is one of his most prized possessions.

It was thus a special moment for both men when Bob finally got to see Keith using his invention live before a huge rock audience. The occasion was a concert on ELP's first American tour held at Gaelic Park, a soccer field at the northern tip of Manhattan. It was a bizarre setting, with an elevated track and subway cars looping around the stadium. ELP, in the early days, attracted the same audience of young white men who came to see groups like Led Zeppelin, Cream, or The Who—an audience brought up on guitar heroes. Keith:

> It was the first time he had ever seen the band live and I remember looking at Bob while I was doing something like "Hoedown" or whatever it was, and he was just standing behind the Marshall speakers and just laughing his head off. And at the end of the show he came up to me and he said, "Man that was incredible!" He said, "The confidence that you showed when you walked up to that instrument and the first thing you played you just hit that note and it was spot on." I felt like saying, "Bob, you don't know—I play this on a wing and a prayer every night."

For Keith, using the Moog before a huge audience was a risk; a keyboard hero needs to be sure that his equipment won't let him down. By taking that risk before proper performance synthesizers existed, he was now reaping the rewards of doing something "which you were not supposed to do."

Bob also remembers that night at Gaelic Park; the limo ride to the concert was not his usual style:

> So here was this soccer field, and there was a stage at one end no seats, there were ten thousand young men it looked like, I don't remember seeing any young women there, throwing their shirts in the air and screaming and yelling at Keith . . . and Keith would have a knife and throw it at his Hammond . . . At the very back of the field there was a series of Porta Johns . . . I was standing back there and I saw Gershon Kingsley . . . there were the doors of the Porta Johns slapping back and forth from one end to the other and

LIVE!

Emerson, Lake and Palmer were holding forth and Gershon was out of his mind he said: "This is the end of the world!"

Bob thinks that Keith's special talent with the Moog was his ability to turn a new sound into a musical possibility. All too often musicians are overwhelmed by the permissiveness of a modular synthesizer. Not so Keith. The pre-set devices on his machine almost certainly helped here. By initially having a limited number of sounds to play with, Keith could make the most of those sounds.

The relationship between Keith and Bob was also important for the development of the synthesizer. The need to build better oscillators was brought home to Bob by the difficulties he saw Keith experiencing. Keith's high-profile use of the Moog also helped generate a huge interest in Moog's performance instruments, especially the Minimoog, and the company's publicity brochures started to use photographs of Keith playing the Minimoog. Keith himself started using a Minimoog in his act: "I'd sometimes use it as a soloing instrument . . . it was very useful for me because Greg was the bass player, he also is a guitar player as well and when he put the bass down we needed to back that up so the Minimoog had a very good resonant thing."

As Keith became anything but the "musician from England no-one had heard of," the Moog company started to give more and more attention to his needs, especially when he toured the states. Keith was known to phone up Bob for technical advice from around the world if something went wrong with his Moog. He would be given access to the latest prototype equipment such as the Constellation synthesizer he once took on tour. This road testing of the equipment also helped the company. As Keith added more and more equipment, it became apparent that he would need his own technical support to help him keep it all running. Rich Walborn, an engineer, was sent from the Moog factory, on Keith's expense account, to accompany him on tour. As Jim Scott put it, "We all looked at each other and Rich was the only guy not married."

The "show that never ends" seemed to have come to an end in 1978 when, after a financially ruinous tour with their own orchestra, ELP disbanded. But in 1992 ELP reformed for yet another sell-out world tour — including five shows at the Royal Albert Hall, London. With the analog revival and interest in the history of the synthesizer increasing in recent years, Bob and Keith are seeing more of each other. Keith still has his "Monster Moog" synthesizer and still occasionally takes it out on the road. Although it was the center of much mayhem and pyrotechnics, it was the one instrument that he was careful never to throw his knife into.

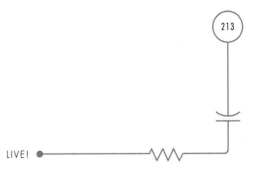

LIVE!

11

Hard-Wired—the Minimoog

I built the Minimoog in my lunchtimes from junk I found in the attic at Trumansburg.

BILL HEMSATH

THE MINIMOOG WAS the first synthesizer ever to become a "classic." Over 12,000 were made, and it was in continuous production for thirteen years. The instrument's portability, ease of use, and relatively stable oscillators made it ideal for live performance. But it also retained some of the instability and rich peculiarities of the analog world that contributed to its fabled sound. In its handsome walnut case and with its unique flip-up design, it became an instrument you could become fond of, an instrument that you could at last really play.

The Minimoog was also the first synthesizer to be sold in retail music stores. Since it was one of the first affordable synthesizers, it introduced many new musicians to the instrument, most notably rock musicians. Some of the definitive rock music of the seventies was played on the Minimoog.

Important inventions are seldom the work of one person. The Minimoog was such a team effort. In the early days of R. A. Moog Co., Bob Moog was *the* engineer. But by 1969, riding high on the success of *Switched-On Bach*, he employed as many as 42 people, including a whole team of engineers. The invention of the Minimoog came at a time when Bob himself was on the road a lot—either giving lectures and demonstrations or desperately trying to find investors to increase the capital base of his company. Indeed, the story of the Minimoog is one of a dramatic struggle as the engineers strived to bring the instrument to market in time to save the company.

Two engineers, Bill Hemsath and Jim Scott, played an important role. Hemsath built the first prototype Minimoog and worked on many of the final circuits with Bob. Jim Scott also worked on many critical circuits and eventually headed the production team. Other engineers played a smaller role. Gene Zumchak, who was Moog's first staff engineer, was an advocate for the Minimoog but left the company before the project was completed. Chad Hunt worked on some of the circuits. Cataloguing everyone's role and finding who exactly did what is impossible. The Minimoog was developed from first prototype to production in less than a year, and everyone at Bob's funky factory chipped in with advice, ideas, and criticisms including the musicians, business managers, and people from the woodshop and even the company's buyer. In short, the Minimoog really was a team effort and in this regard paved the way for later synthesizers that were increasingly built in teams.

◯

Bill Hemsath

Bill Hemsath, joined the Moog company in the middle of 1969. He got his start in electronics early in life—he could solder plugs by the age of three. After he found his father's old ham radio equipment in his grandmother's attic, there was no stopping him. In first grade he was in trouble with his teachers for copying circuit designs out of library books, and by third grade he had borrowed a whole correspondence course in radio physics from a cousin. Hemsath also loves music; he is a self-taught keyboardist. At Case University he studied electrical engineering and took a course in the history of music. The professor, Ray Wilding-White, on learning of Hemsath's engineering skills, invited him to help build an electronic music studio.

They toured the northeast together and visited other studios, looking for ideas. Their itinerary took them to Trumansburg, where Hemsath stayed with Bob Moog in his farmhouse. As an electrical engineer interested in music, Hamsath was thrilled by what Bob was doing.

Back in Cleveland, Hemsath continued to pursue his new interest in electronic music, working with composer Don Erb at the Cleveland Institute of Music. Erb had a small Moog modular system (which Bill helped install), and they put on a series of concerts featuring performances by John Cage and Pauline Oliveros. After being invited to perform with Erb at the Montreal World's Trade Fair, Bill again called on Moog to borrow a prototype polyphonic synthesizer. It seemed that you couldn't keep Hemsath away from Trumansburg, as he returned yet again to use Moog's studio equipment to finish off a recording with Erb. Bob, by this time, had had plenty of opportunity to be impressed by Hemsath's abilities, and a job at the Moog factory loomed: "And we did the session and packing up, getting ready to go, and Moog took me aside and said, 'Would you like to work here?' And I said, 'Yup!'" Bill Hemsath was hired as engineer-in-chief and joined a week after his twentieth-sixth birthday, having served his draft deferral at Cleveland Ordinance, building torpedoes.

○

Jim Scott

Jim Scott (known fondly to his fellow engineers as Scotty) studied chemistry at Berkeley. Without completing his degree, he left for the navy, where he learned electronics working with sonar and gunfire control. After his service he used his veteran's benefits to return to Berkeley to study electrical engineering. His love for music drew him to musical acoustics, and it was a short step from there to electronic music. That step came when Jim heard *Switched-On Bach* for the first time. He immediately wrote to Bob asking if he had any jobs. "Didn't hear anything for a long time. And then one day I got a call . . . and he said that in two weeks there was an Audio Engineering convention in Los Angeles . . . So I went down there . . . and saw the synthesizer for the first time . . . And we sat down someplace side-by-side, and no one says anything for awhile, and he doesn't know what to say . . . I'm kind of embarrassed. He says, 'Well, how about $8000 a year?' I said, 'Okay.'" Jim Scott was twenty-seven years old when he arrived in Trumansburg.

One of Jim Scott's first jobs was to help Bill Hemsath work on the synthesizers being prepared for the MOMA live concert. The roots of the Minimoog can be traced to this concert, because this was the first time Bob's engineers had had to think about what sounds were needed for live performance. When one of the specially modified Moogs was sold to Keith Emerson, it fell to Jim Scott to describe a standard patch. Jim turned to local musicians Chris Swansen and David Borden for help: "We'd connect oscillators up to one envelope generator to a voltage-controlled amplifier to the opposing filter, you know, and that was just kind of a standard block, and you would often just use that basic patch over and over again . . . this is the first step towards the Minimoog."

◯

From the Graveyard to the Min

The components of what made a usable patch were known to most of the engineers and musicians, since they were forever having to demonstrate the Moog. It was Bill Hemsath who decided, after doing this dozens of times, that it was time to build a simplified device based on this standard patch. "I got into a set routine. I would always use the same patching and I would always use the same settings. I would always play the same and so forth . . . And I got to thinking that I would like to have something like that just to tinker with on my own . . . something to play with."

Hemsath's office was in the attic of the former furniture store. The rapid post–S-OB expansion meant that Moog had to open up extra space in the attic, half of which remained unfinished and full of old abandoned junk: "You could walk down this path, and this was Moog's graveyard, where bro-

217

ken cases are . . . and burned out this and burned out that . . . and I would go through that every once in a while in my lunch hour because I'm a pearl diver, and the thought started to gel . . . that there's a half case over there and all I have to do is saw the end off and glue." As Hemsath walked through the "graveyard," he started to piece together what he would need to recreate the most familiar Moog sounds: oscillators, a filter, envelope generators, and a keyboard. Hemsath:

> I loaded up my bottom desk drawer with cast off pieces . . . at noon I'd open the drawer and chomp on an apple while putting stuff together . . . The only piece in there that was a new part—I actually went down to the construction area and stole a new 911 envelope generator. And I think I had two of them and I wedged them back to back . . . Nothing was patched . . . everything was done behind. And the only thing that was adjustable was the knobs . . . I found an old keyboard . . . I was able to slice that down to two and a half or three octaves.

By the time his lunch-time excursions had ended, Hemsath had collected most of the elements of what was to become the Minimoog. He had made the key decision to get rid of the patch cords and hardwire all the modules together in the same box with a keyboard.

It took Hemsath about two months of lunch times to put together his remarkable invention. He remembers completing it by Thanksgiving of 1969: "I made up a little nameplate for it. I thought it should be called the Min, so it said Model A Min, R. A. Moog Company, Trumansburg, New York, and that was stuck on the back." The prefix Min was meant to reflect the smaller size of the instrument. Eventually the name was changed to the more evocative Minimoog (first separated as two words, "Mini Moog," then "Mini-Moog," and finally run into each other in the now standard "Minimoog"). Hemsath himself always preferred the name "Min" for his "puppy," as he proudly referred to it.

Figure 35. Model A Min Moog

Of particular interest is how Hemsath came to place the slide pot used for pitch bending and vibrato to the left of the keyboard—this slide pot later turned into the famous pitch wheel that was to become a standard feature not only of the Minimoog but also of most keyboard synthesizers: "In the original keyboards on the left he [Moog] had a little control panel, maybe three by four inches . . . all I was able to salvage was the left and right cheeks, and the one had a large notch in it. So that's where I put the little slide pot . . . It's playable but it takes a technique." In other words, the famous pitch wheel's eventual location was very much dictated by what was doable and usable. Throughout the development of the Minimoog, contingency would turn out, again and again, to be the mother of invention.

Having completed his device, Hemsath took it home to play with: "I had it at home for probably a couple of months. I played with it every night the first week, and two months later I brought it back to work because it's in the way." For Hemsath, at the time, the box was not particularly significant.

HARD-WIRED—THE MINIMOOG

This was just one of many things he was working on. "I was wading in synthesizers daily, so this is not anything memorable at all." Bob Moog put it pithily when we asked him about the significance of Hemsath's first model: "Well, Hemsath was always doing shit like that. And so was I. You know, we were always putting stuff together. Nothing unusual about it. And on top of that, things are really crazy, money going in, money going out. I was a rotten manager, we had no controls. I didn't have too much time to think about doing stuff like that. I had to worry about was the IRS guy going to come in."

Moog was not that enthusiastic about this particular project—at least not initially. As Jim Scott recalls, "He was kind of letting it happen . . . he couldn't see where it was going . . . nobody was going to buy these things . . . so what are we doing this for? And none of us really knew the market very well, none of us. We were a bunch of engineers, right, theorizing about what the world out there wanted to buy." Moog himself recalls that he initially conceived of the Minimoog not as a performance instrument but as something that would be useful to session musicians who were having to haul their modular Moogs from gig to gig. He also felt that the Min A would have been a better design if Hemsath had been more open about what he was doing.

David Borden, who was around at the factory when the Minimoog was developed and was its first user, knew both Hemsath and Moog well: "What it was was, Moog had the idea for the Minimoog, and he knew how to work it, and he told them that maybe they should build one, but then he wasn't enthusiastic about it afterwards, cause he wanted something that you could screw around with, and the Minimoog was more like . . . plastic plane models that you just fit together."

The ambience that Bob Moog had helped foster in his funky factory encouraged engineers like Hemsath to take the initiative. Bob knew how important tinkering was, even if he was not always enamored with where the tinkering led. Jim Scott recalls those heady days: "Somebody'd get an idea for a new module and they would just go design it. You didn't have to ask permission or anything. That's how the Minimoog got started."

The Min Moog might have remained as oddity were it not for David Borden. As a regular at the Trumansburg studio, Borden became friendly with Hemsath, to whom he often turned for help in operating the studio equipment. Borden saw Hemsath's box and started to borrow it for concerts. He borrowed the synthesizer on such a regular basis that the insignia duck of Mother Mallard was plated onto its side, where it remains to this day.

○

The Min B

The engineers slowly realized that the new Min might be a saleable item. And as 1969 turned to 1970, they had an additional incentive: Moog was going broke. After the failure of the *Switched-On* copycat albums, the market for modular systems went flat. Bill Hemsath remembers that in 1970, instead of selling ten a month as they had the previous year, they were down to one a month.[1] "I certainly recall by the middle of the year Moog was gone most of the time, 'looking for investors,' as he put it, and the rest of us were looking at the ceiling and wondering what we could do. And so somewhere, the idea to sell the Minimoog, or to manufacture one, was really internal."

That the company was starting to go under was apparent to everyone. As Jim remembers, "There was a sense of urgency we might not be able to meet the payroll next week . . . Our credit was cut off from all of our suppliers. We were literally digging transistors out of cracks in the floor and testing them to see if they were any good or not." Leah Carpenter, who was Moog's office manager at the time, remembers several times having to go over to the local bank to negotiate a loan to make that week's payroll.

The Trumansburg engineers were a tight-knit bunch; they shared the same space and constantly collaborated over their various projects. Several

221

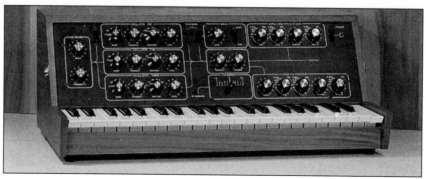

Figure 36. Model B Min Moog

of them had experience in the military, and now they were at last doing what they wanted. They were not going to let the company go down without a fight, and they started to take things into their own hands. Jim Scott remembers that Zumchak kept pushing for a portable synthesizer and that the marketing and business managers, Al Padorr and John Huzar, also agreed with the engineers that "to survive as a company we needed to move into more mass-produced types of things." With the encouragement of his fellow engineers, Hemsath decided it was time to improve the Min. So while Moog was away from the factory traveling, he set about building the Min Model B. The Model A had been "clunky and cumbersome," an old "baroque pipe organ looking thing." He felt the key to winning Moog over was to make it look more slick. He redesigned it to look like a suitcase with four feet on the back and a handle on the front. It was now totally self-contained. It could not only be carried from gig to gig but also easily sent back to Trumansburg for repair.

It was Sun Ra who first saw the potential of the Model B, which he took with him after his famous visit to Trumansburg (everyone recalls that he never paid for the instrument).[2] Jon Weiss remembers going down to New York City to see how the Model B was bearing up under Sun Ra's use. What he saw amazed him: "I happened to hear this machine, and he had taken this synthesizer and I don't know what he had done to it, but he made sounds like you had never heard in your life, I mean just total inharmonic distortion all over the place, oscillators weren't oscillating any more, nothing was working but it was fabulous."

Sun Ra had taken the Minimoog and was making music with it. But there was also something deeper going on: Sun Ra's machine was also not "working" as it was supposed to. This, for Jon, made the Minimoog a musical *instrument*, as opposed to a *machine*. All the best instruments in some sense do not "work" as they are supposed to. It is the departures from theoretical models of instruments—the unexpected resonances and the like—that make an instrument particularly valued.

Many analog synthesists appreciate the subtle complexities that arise from things working not quite as they are supposed to. For example, Brian Eno had to leave a little note on his VCS3 synthesizer telling his technician, "Don't service this part. Don't change this"—he preferred the sound the ring modulator produced when it was "broken." The genius of the Minimoog was that it worked well enough to be a usable instrument in performance but it also left the musician enough room to do things with it that the engineers just could not understand. In short, the Minimoog's fabled analog sound was starting to emerge.

The practicality of the instrument was also something Hemsath noted with satisfaction. When Sun Ra's Model B finally broke down completely, "he simply took it to the airport, bought it a ticket, put it on the plane. It showed up at Mohawk Airlines in Ithaca, New York one day . . . a suitcase, a walnut suitcase. And they looked on the bottom and saw Trumansburg, Moog and they called up on the phone and said, 'We've got this box'—'Yeah, send it over, it's ours.'"

Hemsath's clever design meant that the case fitted over the synthesizer and latched onto it in the same way that a sewing-machine case fits. Hemsath gave Art Phelps, who ran the woodshop, a drawing, and he came up with the details of the case design. The other major change from the

223

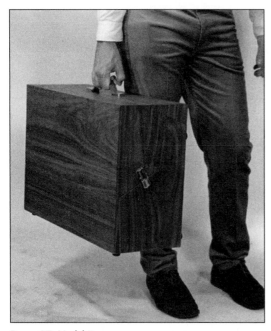

Figure 37. Model B case

Model A was that, rather than use a combination of modules, a whole new front panel was specially designed. The engineers referred to this new design as "the integrated synthesizer" because it no longer used modules. With two half-inch pieces of plankwood at the back, Hemsath made a rack and was able to stack up and connect off-the-shelf cards from standard modules. The decision was also taken to move from two oscillators to three.

It was Hemsath who designed the front panel: "One Tuesday morning before I got out of bed . . . I took a sketch pad and I drew the front panel, pretty much to scale, and I put the little round-cornered boxes around this and that and railroad tracks, as I called them. And that design stuck with synthesizers for probably a decade. Anybody had a synthesizer had to look like that because that's what a synthesizer looked like."

The front panel layout has long been seen as a key element in the success of the Minimoog. Jan Hammer, the Czech jazz pianist who in the 1970s used the Minimoog in the Mahavishnu Orchestra to much acclaim (and later made the music for the TV show *Miami Vice*), is typical of many jazz/rock musicians who took up the Minimoog: "I could not believe why anybody did not think about it sooner. It is just such a straightforward approach. I learned everything I know from the Mini. I never studied synthesizer . . . It was so logical."[3]

Not only is the layout straightforward, but also the knobs and switches are convenient sizes and are comfortably located. The sources of sound—the oscillators—are located on the left and everything flows from left to right, first through the envelope generator, then to a voltage-controlled amplifier, and then to the filter. Switches allow the signal to be routed to the different modulators. Hemsath was acutely aware of design details and what musicians liked to use. As he told us, "A knob is a useful musical instrument." The large knobs on the oscillators made them easier to tune.

The lack of uniformity in the layout is also important: "It turned out that having things not all in military formation made it a lot easier for someone to find a control." Hemsath also chose rocker switches because "they looked nice, but they're tactile, that is, you can feel your way around." Hemsath's design philosophy was to keep everything neat and easy to use. One reason musicians love the Minimoog is because it is simple to operate, and under the pressure of live performance on dimly-lit stages, they can feel easily where the different controls are located.

The game plan had always been to design a number of prototypes, each getting closer to the final product: the Model C would be the production prototype, a Model D would be the "Beta Release" (a model that musicians would be given for final testing); and a Model E would be the final production model. As it turned out, this plan was taken over by events.

◎

The Model C

The success of the Model B, which resulted from Hemsath at last getting feedback from his fellow engineers, persuaded Moog to back the project. "He finally got behind it, said yes, this is a good idea. So we hired a case design team." One of the most distinctive features of the Minimoog is the hinged case that allows it to be carried flat but flipped up for performance. The design team came up with more elaborate designs than the final version. Bob Moog: "They came up with drawings for some very sleek packages indeed—with esculptured plastic cabinets that suggested computer terminals, gleaming multicolored panels, and strikingly shaped controls. We then polled our musician friends to see which designs they liked. Were we in for a surprise! Nearly everybody shot down the sculptured plastic in favor of natural wood and simple lines. We simplified one of the designer's concepts to the point where we could actually make the cabinets in our own modest wood shop."[4]

The unique flip up design was registered, which means that no other company could copy this exact design. Hemsath: "Moog and I went in on Saturday, went down to the woodshop and began sawing wood until we wound up with that." The cases were made of walnut, the same wood used for the modular systems. "Moog had gotten a deal someplace and had a whole barn full of walnut . . . We were going broke at this time—we didn't have any money to buy whatever it was we needed to make cases out of."

The engineers took the original circuit designs used in the modular equipment and tried to improve them. Hemsath and Moog worked on the oscillators to make them less temperature-sensitive. Jim Scott simplified Moog's ladder filter, envelope generator, and voltage-controlled amplifiers to get production costs down. Everything was still hand-wired.

The redesigned oscillators had the advantage of covering the full pitch range in one sweep without having to subdivide the range, as on the modular equipment. Also, with the use of a matched pair of transistors, which were etched together and hence had exactly the same temperature response (Hemsath's idea), they came up with an oscillator that was remarkably stable. Hemsath: "And it worked very well. I remember we'd tune those things, somebody had one in the trunk of his car one winter. He brought it back into the shop. It was stone cold, like 5 Fahrenheit or something. He plonked it down on the bench, and we turned it on and it was in tune, no question. That was pretty good, for those days."

The Minimoog had been designed as a portable instrument, but would it survive going on the road with touring musicians? An opportunity to test-

226

Figure 38. Chris Swansen with his Minimoog

drive the Minimoog C arose when Chris Swansen embarked on a tour of
Europe. He wanted to take the newly developed prototype (one of only
four made) with him. Moog: "Chris had to be out of there by 1:00, and we
got the thing together, sometime mid-morning, at work, and I said, 'Chris,
let me try and drop it, because somebody's gonna drop this.' We dropped it

and it fell apart . . . And somehow we got it back together, and he took it with him . . . That was a good lesson for us in how strong these things have to be."

○

The Pitch Wheel

For many musicians, it is the pitch wheel on the Minimoog that enables them to make the instrument come alive. By bending a pitch or adding vibrato, a note can be given that special personal touch that violinists and guitarists find so important. It was Bill Hemsath, with advice from his fellow engineers including Bob Moog, and in collaboration with a buyer and machinist, Don Pakkala, who invented the pitch wheel. The Model A had used a slide pot on the left end of the keyboard to provide pitch bending or vibrato. Because Hemsath had found this pot (which tended to stick) difficult to operate with the left hand, he had placed it in the middle of the panel above the keyboard on the Model B. Now the slider was about to be replaced with a wheel and moved back to the left of the keyboard.

The sideways mounted wheel with a finger notch cut into it (the pitch wheel) emerged from a completely different project. A customer had ordered a set of six calibrated joy sticks (a means to control two parameters at the same time in the X and Y directions). The project was turning out to be a pain. Hemsath stripped down commercial joy sticks but found he could not eliminate the backlash problem. So he started from square one and built his own joy stick by connecting two pots together with a metal angle bracket. It worked beautifully, with no backlash at all. The cigarette-like joysticks looked like little paddle wheels that could be pushed either forward or backward.

These joysticks were turned into the famous pitch wheels (or "rollers," as Hemsath likes to call them) with the help of Don Pakkala, whose official job at the Moog factory was buyer—he ordered everything from toilet paper to transistors. But in a previous life he had been a machinist, and he

Figure 39. Pitch and modulation wheels of Minimoog

was soon put to work. Pakkala found a way of machining batches of wheels out of perspex plexiglass (later the pitch wheels were injection molded). Hemsath and Pakkala worked together as they refined the new design, adding a notch in the middle and experimenting with springs and later a detent device to ensure that the wheel returned to its middle position. And when it was ready Hemsath and Pakkala proudly showed it around the factory. It was mounted on the left because "I just simply watched a lot of musicians and they could either play the ribbon with their left hand, or anything over here, just fine."

The invention of the pitch wheel is one small part of the Minimoog story, but it has had a lasting impact. Jan Hammer feels that "the pitch wheel really sets the Mini apart from any other keyboard synthesizer."[5] Roger Powell (a well-known mid-1970s synthesist) found that the pitch wheel and the modulation wheel (which sits beside the pitch wheel and allows the degree of modulation to be controlled) "are the most humanized controls that I have found yet on synthesizers."[6]

○

Production

Having built what looked like a viable design, the company now had to make the decision as to whether to put the Minimoog into production and, if so, how many they should make. They faced an uncertain future because until this point no one had tried to sell a portable synthesizer. This had been Moog's worry all along. Jim Scott remembers that the transition from the Model C, the pre-production prototype, to the D, the production model, happened almost by accident and required a small moment of insurrection among the engineers and production managers. Moog was away on another of his speaking tours and had given permission for another ten model Cs to be built by hand-wiring them on perforation board. Scott: "As soon as Moog left we all looked at each other and said, 'You know, if we don't get this thing engineered to the point where it can be produced, we're all going to be out of a job anyway. And it's going to take us forever to hand-wire these things . . .' And so Huzar, I guess it was, and Padorr said, 'Okay, just go ahead and lay out the boards.'" Hemsath laid out two circuit boards, and Scott and Chad Hunt did one circuit board each.

When Moog returned a couple of weeks later he faced a fait accompli. The Model D, which was a Model C with circuit boards, had been built and was in production. Jim Scott takes up the story: "And he came back and found ten D Models nearing completion, and he was not pleased. And he called us all into his office, and he let us know in basic Anglo-Saxon exactly what he thought of all of us . . . And we all just kind of sat there and looked at our fingernails and nobody said anything."

Bob Moog can't remember this particular meeting, but it was likely that he was indeed "pissed off." From his point of view, his young and inexperienced engineers would sometimes try to design new devices and wanted some, if not all, of the glory to fall to them. "I, on the other hand felt that:

Figure 40. Minimoog Model D

(a) my engineers were just learning to design synthesizers, whereas I had two decades of electronic instrument design experience behind me, and (b) it was my company with my personally-guaranteed debt load, and I had a right to participate in decisions such as when a new product was ready for production."[7]

Some last-minute design decisions were dictated by the financial pressure on the company. For instance, they had originally planned to use five different colored rocker switches, with different colors denoting the different functions. But the company making the switches heard they were in financial trouble and refused to deliver any more switches. Scott: "So we went around and changed our idea about what colors, which went where. And that's what stayed with it forever."

The original plan to build a final production model—the Model E—was never fulfilled: "It was manufacturable, as it stood. And so we made a few more and a few more." Scott eventually wrote a production manual, and one of the most successful synthesizers in history stuttered off the production line. The retail price was set at $1,195 dollars.

The Minimoog had been developed under extraordinary circumstances

Figure 41. Minimoog in production

in less than a year. Hemsath and Moog introduced it at the October 1970 meeting of the AES. The time for the portable synth had finally come: at the same meeting were displayed the Putney (the American version of the VCS3) and ARP's 2600. Dick Hyman, a high-profile Moog musician, premiered the Model D in the August 1970 "Jazz in the Garden" series at MOMA.

Bob had no idea how successful the Minimoog would become, "And I remember thinking, and saying to a lot of people, we're going to make a hundred of these and then we'll stop and see where we are. You know, the funny thing is, we never did stop!" The success of the Minimoog, however, came too late to save his company from being taken over. In May 1971 Bill Waytena assumed Moog's debts and moved the company to Williamsville near Buffalo, leaving behind most of the engineering team, whose families were established in the Trumansburg area and who preferred to remain in a rural setting. With their electronic skills, they soon found other jobs. Neither Hemsath nor Scott made the move, although Scott returned to the Moog company a few years later to develop the Micromoog.

The Minimoog itself went from strength to strength. By the early seventies it was so successful that the name "Moog" had become synonymous with the Minimoog. The modular Moog was far too complex to sell through retail stores, and with a glut of modular systems on the market the Moog company put nearly all its effort into advertising the Minimoog. Un-

232

like for the modular Moog, there was even a manual and a book of "sound charts." The latter provided musicians with tear sheets with which to document their own sounds and reproduce the sounds of acoustic instruments, and the most popular ELP and Rick Wakeman sounds. It was not yet the pre-set synthesizer of the digital age, but it was a step in that direction.

Some musicians already familiar with the modular systems and their greater flexibility noticed the loss in the range of sounds available on the Minimoog. David Borden used both: "If you had a certain sound that you made with a modular synth you couldn't get it with a Minimoog. And actually, Steve [Drews, of Mother Mallard] didn't like working with the Minimoog that much, except for certain stock kinds of sounds . . . for his real unique sounds he used the mod-

Figure 42. Bob Moog at Minimoog, 1970

ular synthesizer." Bernie Krause noticed a similar loss. For the serious synthesist wanting to make studio electronic music, the Minimoog offered far fewer possibilities.

But for the gigging musician wanting something that was easy to use in live performance, the Minimoog was a killer synthesizer. And great musicians like Sun Ra could always do the impossible anyway. As Don Preston points out, "I did a solo on one of the Mothers' albums and Paul Beaver played it for Bob Moog, and he listened to it and he said, 'That's impossible, you can't do that on a Minimoog.' I always enjoyed that comment."[8]

The Sound of the Minimoog

Some instruments acquire a legendary status. The sound of the Minimoog is for many the definitive analog sound. Old Minimoogs command top prices in the vintage synthesizer market. For many musicians, its bass sound has not been bettered even on more modern digital synthesizers. The Minimoog, as only a classic instrument can, has a special mystique. Is it the walnut case? After the initial supply of walnut was exhausted, later ones were made out of a wood that was stained to look like walnut. Jim Scott recalls that the real walnut models were preferred by musicians, as if somehow the wood itself could provide the magic sound. Is it the Trumansburg ambience? David Borden is convinced the earlier Minimoogs must have been better because the team in Trumansburg really cared about the instruments. Different numbered sequences of models are discussed by enthusiasts in order to elicit clues as to which ones have the best sound.

Figure 43. Moog Music logo

Bob and his engineers are rather skeptical about claims to the superiority of the earlier models, especially as the oscillators on the Minimoog were improved over time. And as to that special analog sound, both Scott and Hemsath think they can explain part of it. One of the key features of the Minimoog, which differentiates it from other synthesizers, is that the oscillators do not lock together. If you tune them together, they stay slightly out of tune with each other and roll through each other, producing a very pleasant choral effect. It was another accident that produced this feature:

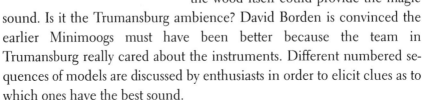

the Minimoog's power supply, as Jim Scott later discovered, was noisy enough to jar any locking tendencies. When he redesigned the power supply to try and make it less noisy, much to his surprise, the oscillators now locked.

What makes for the good sound is still something that is laden with mystery. For Scott, it is the fact that the circuits were deliberately overdriven, producing modulations and distortion: "It was something like vacuum tubes, in that the circuitry would not suddenly go into clipping, it would distort gracefully . . . Also, the circuitry was inherently wide band . . . It passed frequencies far beyond the audio range . . . And we're getting into guess work here, but the feeling is that there were things that happened up in the ultrasonic range that can cause inner modulation and distortions, [this] reflects back and can be heard in the audible range."

In the absence of any certain theoretical understanding of what exactly makes good sound, the engineers found that they relied upon their own ears. Scott: "The reason it sounded good was because none of us were using very much in the way of instrumentation. We were using our ears to set parameters . . . If it didn't sound good, you kept fooling with it until it sounded right." Watching and interacting with musicians was also crucially important. Hemsath and Scott were both close to musicians such as Chris Swansen and David Borden and would help them set up their equipment and build special devices for them. Jim Scott maintains that all the Moog engineers were "frustrated musicians."

And finally, there was teamwork in building the Minimoog. Bob Moog himself recalls that "I remember a lot of talk among all the engineers, to the point that it was impossible to say whose idea was whose. I remember talking about left-hand controller wheels, three oscillators, keyboard length, etc."[9] Hemsath sees it like this: "I was just there at the time. I just think of it that I got to drive the bus for a little while, and then somebody else did. It was fun. But I've been working now 35 years, and it only took me six months to do that one."

The tragedy for Moog personally was that the Minimoog came out too late for him to keep control of his company. The funky factory that had changed the face of popular music and the gang of engineers who worked for the love of it were soon going to be dissolved. The soulless suits were about to take over and shift the factory into a grim ex-gelatin plant on the outskirts of polluted industrial Buffalo. Bob's rural bliss was over, as were the halcyon days of the Trumansburg factory.

Although Bob Moog did not have as large a part to play in the development of the Minimoog as he had had in the development of the modular system, the Moog way of doing things was responsible for so much of the success. The strategies he had set in place earlier eventually carried him along. The engineers he hired were all the sort who liked to tinker and make do, exactly as he had done earlier. The space he provided them to work in ensured maximum interaction and encouraged them to be as inventive as possible. The studio musicians he employed, and the other musicians with whom he worked, were constantly interacting with his engineers. And he had, at last, found the ideal product to mass produce—a reliable and portable instrument that had the sounds and controls the musicians wanted.

As the engineers took things into their own hands and pressured him to go straight into mass production, Bob was naturally cautious. He had, after all, seen interest in his modular synthesizers wax and wane. He no doubt also recalled the nightmare of mass-producing guitar amplifiers. Bob had no way of knowing that the key action was now switching away from Trumansburg and out into the world, where a new group of young musicians were starting to realize what his marvelous new instrument could do for them. The man who first saw the new dawn was, appropriately enough, an ex-evangelist preacher, David Van Koevering.

12

Inventing the Market

I don't think I, or anybody else in the company, went into a music
store before . . . March of 1971.
BOB MOOG

IT WAS 1970 and Moog's small Trumansburg factory was
about to witness one of the biggest transformations the synthesizer industry has undergone. Moog and his loyal team of engineers
were at first only dimly aware of what was happening. Their struggling
company, in desperate financial straits and under much pressure, had
brought out the Minimoog—their first portable keyboard synthesizer. It
was set to become a classic but they, as yet, had no clue who would buy this
synthesizer or indeed whether it would sell at all. Even Moog himself
couldn't fathom who the customers might be.

Minimoog production was started in batches, just enough to meet current orders. Then something happened. One salesman started to buy more
and more Minimoogs; furthermore he was selling them in retail music
stores. That salesman was David Van Koevering.

David Van Koevering was a man with a vision. He heard the Moog synthesizer and it changed his life. He still remembers that moment. It was after he had first met Bob Moog and they had gone to a concert featuring
Gershon Kingsley's "First Moog Quartet" at Carnegie Hall: "I saw something . . . the power of the sound, the sonic energy, and I believed that it

could become common, and I imagined it as powerful as the electric guitar to the first guys that ever played this thing . . . And I argued with Bob that it is a performance instrument."

◯

Little David

David Van Koevering likes to perform. He has been doing it since childhood, when he won prizes for playing the ukulele behind his back (long before Jimi Hendrix discovered this performance technique). By the age of nine his father had taught him to play over fourteen different musical instruments. He performed with his father in churches from Michigan to Florida as part of a novelty musical instrument show that combined a religious message with music. When the show was later covered on TV, David Van Koevering became known as "Little David." The show included Swiss hand bells, frying pans (tuned by denting them), gear wheels, musical stones (cut from rock beds in Kentucky and played by hammers), and much more.

Eventually, Little David took over the show from his father and refocused it on education, explaining to school kids some of the fundamental principles of musical instruments. He added electronic instruments, including a theremin. It was no doubt the theremin that sparked Bob's curiosity and drew him to attend one of Van Koevering's shows at a school out on Long Island. That same evening Bob was going on to Gershon Kingsley's concert in Carnegie Hall, and he invited Van Koevering to come along.

Van Koevering drove back to Trumansburg with Bob, and his vision of the Moog as a performance instrument was soon put to the test. Neither Bob nor his engineers thought that the modular Moog could be used for routine live performances. Van Koevering acknowledged the limitations: "It didn't stay in tune, it was a tough thing to set a pre-set up on . . . and everything you did with every one of the knobs gave you a brand-new sonic experience, some of which you never could repeat again." But when you've

238

got a vision and you're on a mission, you are hardly going to let mere details stand in your way. Van Koevering left Trumansburg armed with the latest weapon in his sonic arsenal—a modular Moog synthesizer. He was determined to take the Moog on the road and bring it to the people. He added it to his show, with the word "MOOG" displayed in large letters, and discovered, as he had expected, that his audiences loved the new instrument.

○

From Swiss Bells to Taco Bell

Buying a modular Moog synthesizer (albeit one of the smaller models) was a big investment for Van Koevering. But he soon hit upon an ingenious way of making his investment pay off. He contacted one of his Florida business acquaintances, Glen Bell, the founder of Taco Bell. At this time Glen Bell was trying to get the taco, a southwestern form of food, established in the southeastern part of America. He was opening restaurants everywhere in the southeast, and he was looking for ways to introduce families unfamiliar with tacos to this new dining experience. David Van Koevering,

Figure 44. Taco Bell promotion of Moog, 1970

on the other hand, was looking for ways to introduce people unfamiliar with the Moog to this new sonic experience. Moogs and tacos became linked.

The deal was this: after each school show, Van Koevering gave out free family coupons for tacos to be redeemed at the area's newly opened Taco Bell restaurant, where he would be playing that evening. Van Koevering was very happy with this arrangement (for which he was paid handsomely), and tacos have never tasted as good. The spread of tacos, like the Moog itself, was the latest hip thing to hit the sunshine state.

○

The Island of Electronicus

Van Koevering could see the impact the Moog was having, but he wanted to do better than just make music to eat tacos by. He set out on a much more ambitious project: to develop a totally new sort of electronic event where the full sonic power of the Moog synthesizer could be experienced. He describes his inspiration as a vision: "And I saw something . . . I saw a room. I saw this huge ceiling, I saw this glass dome at the top, and I realized that the room existed, that I'd seen it in a newspaper story about an island off from St. Petersburg, Florida, called Tierra Verde . . . I could see a show, it was in me, it was all over me. I saw speakers hanging on all the beams."

Anyone who has met David Van Koevering will quickly realize that he is the sort of guy who can sell anything. He can giftwrap any object in a compelling verbal spiel that makes it irresistible. On one famous occasion, later in his life, he took a whole bunch of Memorymoog synthesizers that the Moog company could not sell, repackaged them as the "Sanctuary Synthesizer," and sold the complete stock to American churches! Salesman or visionary, David Van Koevering was extraordinarily effective at what he did.

It so happened that the island Van Koevering had dreamed of was leased by his business partner, Glen Bell. He located his "Island of Electronicus" on this artificial land mass connected by a causeway to the Florida coast. He designed a "Happening Stage" for the domed auditorium with "LOVE" and "PEACE" emblazoned as a backdrop. He had the audience seated on pillows, added a light show and a massive sound system, and designed posters that advertised shows with "Switched-On Sounds Moognifisant" provided by two Moog synthesizers "played live on stage." Tickets cost $3.

Figure 45. The Island of Electronicus: David Van Koevering (2nd from left), Bob Moog (seated) on stage

The Moog synthesizer was at the center of Van Koevering's presentation. Radio commercials for the Island ran as follows:

> *What* is the Moog Synthesizer? The Moog synthesizer, an electronic instrument capable of producing any sound imaginable, that will play with your mind, with your body and heighten the horizons of your soul. Come to the Island of Electronicus and . . . experience the sound.
>
> *When* is the Moog Synthesizer? The Moog synthesizer is NOW! It brings your mind, body, and soul together in an im-

INVENTING THE MARKET

mense, creative explosion of thought and feeling. Come out to the Island of Electronicus . . . and share in this vast creative force.

Where is the Moog Synthesizer? The Moog synthesizer is here at the Island of Electronicus, on Tierra Verde. It's here *now*, with a new concept in sound. It's here to stimulate your feelings, thoughts, and your love for your fellow man. Come to the Island . . . to experience this sound, share your creativity and share your love.

No doubt with their curiosity piqued about a mere machine that could deliver all of the above, people started to flock to Van Koevering's shows. "And the place filled up. Three dollars a head and they're lined up, the parking lot is full." A large proportion of the audience was teenagers. There was no liquor license, but he did serve food. On nearby Treasure Island was Lenny Dee with a Hammond organ show. "Lenny Dee showed up [and] said, 'If you ever serve liquor here . . . I'm out of business.'"[1]

Before David launched his technological assault on the audience, he gave them his preacher's rap. As the lights dimmed his voice came over the PA: "Tonight we take you one step closer. All the sounds you've ever heard are like a second. The Moog is an eternity . . . Tonight we witness the dawn of a new enlightenment . . . seclude yourself now and let the music sweep you away and into the dawn. Seek to become newly aware of yourself, the world of nature around you, the people near you. And if you feel it, express yourself."

Then he would launch into the first piece, "Dawn." Inspired by a Beaver and Krause–style combination of sound effects (birds and insects) and music, slowly the Moog cast its first sonic rays upon the planet. Part of the show was creating soundscapes:

And we'd start a motorcycle up—you'd hear a Minimoog sound like a motorcycle, you'd hear 'em kick it over, and then we'd take noise, and you'd hear 'em choke it . . . and you could hear that filter screech and like a wheel would chirp . . . with the Doppler effect . . . we'd have this motorcycle flying around the room . . . now, we did this with two Minimoogs—a four-cylinder sports car would start its engine . . . And you'd hear the motorcycle going one way and you'd hear the sports car go the other way, and a horrendous crash would happen over the stage and parts were rolling all over the room. And the audience would go nuts. They'd stand and they'd cheer and they'd clap.

Part novelty show, part happening, part concert, part *son et lumière*, part a revivalist meeting—it had it all. Although this venue and audience were a world removed from Haight-Ashbury in the psychedelic sixties, the use of electronic music and light shows to induce transcendent states was similar. It really was what the Trips Festival had advertised—an electronic experience without LSD. Van Koevering wanted his audience to experience a Moog-induced state of reverie.

Van Koevering used six of the first Minimoogs ever produced as part of his show. He encouraged audience participation, and he had some of the Minimoogs set up at the front of the "Happening Stage" for people in the audience to play with. He soon found that members of local rock bands were coming to the Island just to play the Minimoogs: "Members of rock groups would come sit in the pit . . . there was a place you could sit that had a Minimoog . . . you could put headsets on and you could play your Minimoog along with the show . . . and with a switch I could listen to this Minimoog and this Minimoog, and these kids were good, some of these kids were great, and I could patch them into the sound system . . . a spotlight would come on over the kid, over that pit, and you could hear the kid play along with us."

The Island of Electronicus was a huge success. Bob Moog himself came down for a special event with live radio coverage. Van Koevering started to employ other musicians who would use the Moogs to make advertising jin-

243

gles during the day. But eventually Van Koevering decided to give the island up. He was tiring of the shows (three a night, five days a week, and a special show for churches on Sundays), and he had an even bigger project in mind. He was going to sell Minimoogs: "And I said, 'I'm taking the Minimoogs on the road and I'm going to establish a dealer network.' And my wife thought I was absolutely insane. Bob Moog told me they weren't going to sell. You can't sell this in music stores."

Van Koevering likes to boast about how many music stores he has been thrown out of: "I've been thrown out of more music stores than any man alive . . . and I had to invent the market." To help him invent the market, he first set up a company called VAKO (after VAn KOevering) Synthesizers with a partner, Les Trubey, who ran a music store in St. Petersburg, Florida. Trubey could see the demand for Minimoogs in the St. Petersburg area stemming from the Island of Electronicus. The plan was that Trubey, through VAKO, would buy the Minimoogs direct from Trumansburg, and he would then send them on UPS to dealers and stores that ordered them through Van Koevering, who would be out on the road.

○

On the Road Selling Minimoogs

Van Koevering was completely confident that he was up to the challenge: "I owned a brand-new Cadillac, told my wife, I'll be back when the Minimoogs are gone, and I piled all the Minimoogs in the trunk, in the back seat and I'd go to the first city." This was Gainesville, Florida, where there was a big music store, Lipham Music, and the University of Florida: "I go to Buster Lipham . . . and he laughs at me. He says, 'You want me to sell that thing? Show me how to do a violin, show me how to do a flute' . . . And he said, 'If you can prove to me that musicians will do this, you come back . . . and I'll sell them.'" One of the people working in the store, Bob Turner, was a rock musician. He was impressed by Van Koevering's demonstration and said: "'I'd love to play that in my rock band.' And I shoved it across the table at him, I said, 'Well, then, do it!' And he said, 'I can't afford to buy it.' I said, 'Who said buy? I said do.'"

Bob Turner was soon recruited by Van Koevering to be his first sales rep. Van Koevering realized that music stores were unlikely to take this new instrument with its 44 different dials and switches—it was just too unfamiliar. The strategy he hit upon was to go directly to the musicians, persuade them to buy the Minimoog, and then take them back to the store. Then, having seen that customers existed, the store might be persuaded to stock the instrument. He concentrated almost exclusively on rock musicians to begin with; later he would widen his net to include all sorts of gigging musicians.

Whenever Van Koevering entered a new town, the first thing he did was to find out where the clubs were and head there. Next stop on his trip was Jacksonville, Florida. There Van Koevering had a contact, Bill Hoskins, who taught the modular Moog synthesizer at Jacksonville University. With Hoskins's help he got a list of clubs and went out in search of rock musicians. Soon Van Koevering realized the practical difficulty he faced. The rock keyboardists to whom he showed the instrument had never played a monophonic instrument before: "I'd set a synthesizer on top of their Fender Rhodes or on top of the Hammond B3 and they didn't know what it was, and they'd hit it with a chord and nothing would happen." He started to teach them how to play the Minimoog. He devised a very simple way to find sounds—it was actually the same method that his father had used to teach him to play novelty instruments:

> And I carried rolls of tape in my pockets, colored tape, mystic tape, with a scissors, and I would create a sound that he liked, and we'd put red slivers on all the 44 knobs and switch positions that meant all to red is sound red, and all the yellows are sound yellow, and [so on] . . . and those were the pre-sets. "But on the way from

245

one pre-set to the other, anything you find that's musical, play it, experiment, create a song for the sound, create a mood with the Hammond or the Fender Rhodes to accompany this melody line." And I was teaching them synthesis, and they'd get it.

The little slivers of colored tape cut into a V-shape enabled the musicians to quickly move between the sounds. What Van Koevering was doing was making a precursor to the sound charts that Moog produced later to enable musicians to find sounds on the Minimoog.

Once the musicians had mastered the basics, Van Koevering offered to lend them the instrument to try out live. He would only work with a band that had a contract with a club or hotel. It meant they had some money and also that they would be performing more than one set. Before they went on stage, he prepared the synthesizer by pasting "MOOG" in large, silver mailbox letters on the rear of the walnut cabinet. He instructed the band to ask the audience during the first set whether they liked the sound of the Moog. This call and response routine must be one of the oldest in show business, and it worked perfectly. Soon the audience was excited about the new Moog.

With interest in the instrument now at fever pitch, Van Koevering made his move. He did something which, like all great sales ploys, is totally counter-intuitive.[2] He left the club:

> First set's over, I pull the slow-blow fuse and I split. They couldn't start that thing up if they wanted to without that slow-blow fuse in it . . . I've even had kids try to short the suckers out while I was gone . . . now, the audience has been stimulated . . . and a request would come up from the floor, "Play that thing." Now, the manager didn't know that I'd split, and the manager would say to the kid, "Play the Moog." He'd say, "I can't play the Moog." What I'm doing is I'm conditioning the manager to know that this is meaningful for his attendees.

Van Koevering would return at the end of the second or third set. "I'd come back, and the kid's mad, he's been bawled out by his boss, the boss is upset with me, 'Why the hell did you leave? You're not supposed to leave. You get this kid all going and then you leave.' 'I had another show I had to go do at another club.'" Often Van Koevering would just sit in his car outside the club and read a book. Now all he had to do was close the deal:

> The next day, or that night, I would say to the kid, "You ought to have this." "Oh, I know, I've got to have this." "Well, I can't do it again, this is it. You get your loan together . . . You go to your girlfriend, you go to her mother, and you get a loan [using her] signature at a loan company. You call me, I'm staying at this hotel, you do that today."—it's 1:00 in the morning, 2:00 in the morning— "You stay up and you do this before you go to bed!"

Sometimes Van Koevering would have the loan papers ready and waiting after the gig to be signed. This high-pressure sales tactic worked: "I'd get three or four of these kids doing this in a city. There's hundreds of cities that I've done this in. I did this until we had a Moog network selling Moog synthesizers coast to coast."

The introduction of consumer appliances into America was often accompanied by such hard-sell routines.[3] These methods were typically used for a new product being sold to an entirely new group of users. Van Koevering was simply following in the tradition of such itinerant salesmen. From his perspective he was not only selling the young musicians instruments but also giving them the means to release their musical energy: "Bob Moog has respectfully called me a great salesman, and I suppose that that's a correct term, but there's a passion that I carry for creativity and it's my job to unlock that in that kid . . . I take it very personally . . . And if I can do that I can change that kid's life."

Van Koevering was relentless as he traveled from city to city, crisscrossing America. Like all great salesmen, he refused to take no for an answer and would not leave a city until he had sold at least one Minimoog. He was persuading not only musicians to buy the instrument but also dealers to stock the instruments; he was also recruiting sales reps. In short, he was building a market.

But he soon realized he was not alone. ARP too was starting to sell synthesizers in retail music stores (in fact, Van Koevering claims they were following him around). He often found himself giving clinics (demonstration classes) on these other instruments as well. There was a camaraderie among these early pioneering salesmen. Like the engineers and musicians, they too felt the excitement of the new medium.

○

Lucky Man

While Van Koevering was out on the road, the impact of the Moog on popular music was slowly growing. Van Koevering was aware of which Moog records were selling and carried his own copies with him to ply reluctant dealers with. One hit record in particular made a big difference:

> Here's the big breakthrough . . . "Lucky Man" shows up, and then there's a Keith Emerson tour, followed by a Rick Wakeman tour. I mean, we knew where these artists were by the cities that were calling . . . because when the guy did a show they'd go to the music store to find one, and the music stores had to get a Moog—so we knew where Keith was by the phone calls . . . we knew he was in Boston because you got 30 phone calls from Boston. We knew he's over in Wilmington, or he's in New York City or he's in Chicago, or whatever.

The impact of the progressive rock movement, with soloists like Keith Emerson and Rick Wakeman, further served to legitimize the instrument. It was now not only a new instrument but also a cool instrument that rock stars had endorsed. Young rockers could see for themselves the effect

Figure 46. Minimoog sales brochure

Keith Emerson was having on his audience, and they too wanted to become "keyboard heroes." It was exactly the sorts of Moog solo played by Keith Emerson and Rick Wakeman that they wanted to recreate on their own Minimoogs. Part of Van Koevering's pitch was to appeal to the impact they could have as a rock *soloist:* "The Minimoog could make, because of its sonic energy, it could make the keyboard guy a superstar—a monophonic, piercing electronic sound coming out of four or five . . . amplified speaker stacks, could give him some energy and he could compete with the guitar, and he wanted to do that."

Well-known keyboardists like Gary Wright and Jan Hammer adapted their Moog keyboards to wear over their chests like guitars. This development was eventually taken to its logical extension by the Moog company when, in the early eighties, they produced the aptly named *Moog Libera-*

moog liberation

Figure 47. Moog Liberation

tion—a synthesizer keyboard shaped like a guitar that could be worn over the shoulder. This attempt to turn the synthesizer into the guitar shows again the power of that particular cultural icon.

○

Moog Moves and Van Koevering Goes to the World

Van Koevering's success selling Minimoogs did not go unnoticed in Trumansburg. Scott: "All of a sudden we'd get an order from Van Koevering for a hundred or so of these things, whoa!" But by this point Bob had managed to sell the company to a venture capitalist, Bill Waytena. Moog: "He had his group of investors—doctors and lawyers, various businessmen and accountants—and he would make them part of every deal he did. You know, get five thousand from this one, five thousand from that one, get a quarter million dollars to go into something and in two years it would double and these people were happy as clams. So that's how he worked." Waytena paid nothing for the business but subsumed Moog's $250,000 of debts. Waytena's deal was somewhat wistfully presented at the time as a "merger."[4]

Waytena had got interested in Moog through Gene Zumchak, who happened to live near him. Zumchak (known to his fellow Moog engineers as Zummy) was one of the first people to be "let go" by Moog partly as a result of the downturn in business and also because, by all accounts, he was not the strongest member of the engineering team. Zumchak had been an early advocate for the portable synthesizer and had taken his plans to build such an instrument to Waytena (who, like Zumchak, was Ukrainian). Waytena formed a company, Musonics, to produce and market the Sonic V synthesizer (engineered in part by Zumchak).[5] But realizing that he was not going to be able to sell many synthesizers without the name and product experience of Moog, he decided to buy that name. After a period of on-off negotiations, he finally bought the company in spring 1971, moved it to Williamsville near Buffalo, and changed the name to Moog/Musonics (soon the Musonics part was dropped in favor of Moog Music). Bob Moog moved to Williamsville with the company.

Waytena insisted that Moog start attending the National Association of Music Merchants (NAMM) shows. Bob: "As crass and as unmusical as he was, he probably, because he was so far out of the mainstream of all this, he was probably able to look at it and see where it was going in two or three years . . . And what he saw was that in our company in 1971 it was at the cusp, going from one thing into another, and that in a couple of years we were going to be part of the musical instrument business, but we weren't

251

yet." Before Waytena's involvement, Moog had gravitated toward his fellow audio engineers at the biannual AES meetings, where he would also demonstrate his new products. But if the synthesizer was to become a musical instrument rather than a specialized piece of audio hardware, this meant reaching music stores and dealers. And the most effective way to reach such people was through NAMM.

Today synthesizer companies like Roland, Yamaha, and Korg have some of the biggest displays at NAMM shows. There they recruit dealers and show off their latest products. But back in 1971 synthesizer manufacturers had never attended such shows. The Minimoog was first demonstrated at NAMM in Chicago in June 1971 but did not make much of an impact. Von Koevering: "Dealers didn't get it . . . they didn't know how to demonstrate it [and] they couldn't sell it."

Part of the difficulty the Moog company faced at shows was that the music instrument trade was still dominated by organ companies who displayed their latest products with slick demonstrations. ARP first developed the sort of musical demonstration with synthesizers necessary to have an impact at such shows.

Before Moog was taken over, VAKO Synthesizers had always operated as an independent entity. Waytena's plan from the outset, however, was to bring Van Koevering into the fold. Van Koevering, as always, was out on the road when the transfer of ownership occurred. He got a telephone call from Waytena: "He finds me on the road in a music store, and he said, 'If VAKO Synthesizer ever gets another synthesizer you're going to meet me at the airport in Atlanta tomorrow—I just bought the Moog Company.' . . . So I went to Atlanta, and he told me, 'You're going to become vice president of Moog Music.' I said, 'I don't think so.' He said, 'If you ever sell another synthesizer you will.' And I knew that I couldn't argue with him, he had deep financial resources."

Van Koevering had never needed a formal contract with Bob Moog. They had become friends, and anytime David needed synthesizers Bob, of course, was eager to supply them. Now Waytena held the upper hand, and David asked him what he wanted. "He said, 'I want you to come to New York and you're going to give up your network and you're going to sell synthesizers to the world.' And he appealed to that part of me that had this mission. I wanted to go to the world."

Thus it was that David Van Koevering, former novelty instrument showman, became a vice president of the best-known name in the synthesizer business. It truly was a meteoric rise in status and too good an opportunity to miss. As head of Sales and Marketing he was now in a position to recruit a sales force and repeat on a larger canvas what he had already carried out—he was going to sell synthesizers to the world.

He visited the European equivalent of the NAMM show (it's actually bigger), the Frankfurt Music Messe, to put his international sales plan into operation. He recruited dealers who were fully committed to the products and who were prepared to sell them in the way that he knew worked. Van Koevering: "They had to come over here [to Williamsville]. They had to take a tour of the factory, they had to see how a Minimoog was made, they had to learn how to adjust the Minimoog on the inside for calibration reasons, how to put circuit boards in them, how to fix them in the field, and go out on the road with me. Now we're building a network over here, now we've got road men over here traveling—the same things that I did in Florida."

Van Koevering demanded a lot of his dealers, but it worked; and soon the Moog took off in Europe. The list of distributors for Moog performance synthesizers at this time included outlets in Canada, Denmark, England, Finland, France, Germany, Greece, Hong Kong, Italy, Japan, Netherlands, Norway, Portugal, Spain, and Sweden. All in all, 685 different dealers made up this international network. The stepchild of the modular Moog synthesizer that had first emerged from the Trumansburg attic only three years earlier was now on the shelves in all the best music stores everywhere.

Back in the United States, Van Koevering continued to do what he al-

Figure 48. David Van Koevering in Williamsville sales office

ways did—go out on the road and sell synthesizers. Three of his sales books from this period survive and reveal how busy and successful he was. In a nine-month period between December 1971 and September 1972 Van Koevering recorded 121 different sales in 107 different cities in 25 different states. In this period he sold 86 Minimoogs and an additional 168 synthesizers (Sonic V, Sonic VI, and Satellites), plus a range of accessories, including Moog T-Shirts. He set up 47 Moog dealerships and also arranged bookings of special demonstrations to be given by Bob Moog himself (referred to always as "Dr. Moog") and for Moog LPs, literature, tapes of radio and TV shows, and the like to be shipped to many of these stores. Van Koevering was nothing if not thorough. The Authorized Dealer list for

Moog Music when Van Koevering was vice president had entries for 241 dealerships in 42 different states.

○

The Vision and the Mission

David Van Koevering was indeed a man with a vision on a mission. For him, the two things were linked. It is easy dismiss his "visions" as just so much salesman's bluster or hocus pocus. But this is to misunderstand how such charismatic salesmen operate. *All* of Van Koevering's many projects were accompanied by this sort of visionary rhetoric. Whether selling God or synthesizers, or synthesizers for God, he still had to persuade a group of people to commit to something. Building commitment to a product is something all effective salespeople do, whatever their product, and having a vision that can be shared is a compelling way to build such commitment.[6]

His mission developed as his circumstances changed. Like Moog, he probably did not plan it all out in advance. He responded to the situation he faced and was prepared to make changes, uprooting his family if necessary. As a user of the Moog himself, he was able to see directly what worked and what didn't. His introduction of the Moog into Taco Bell, which sounds now like a dead-end, was actually a good way to find out how ordinary people reacted to the instrument. The Island of Electronicus project was a remarkable way to bring to fruition his vision of the sonic power of the Moog—a bit like running a test laboratory for the new instrument. Big companies who make consumer appliances have such in-house test laboratories. In this case Van Koevering built his own laboratory with his own captive audience. It gave him a chance to interact with new users, especially rock musicians, and learn from them what their requirements were. It gave him a chance to see the sales potential of the Minimoog among this new group of users.

Van Koevering's last change of direction—from the lab back out into the world—was the most extraordinary of all. He took the lessons from his labo-

255

ratory and applied them in the real world with great success.[7] In the process, he managed to persuade numerous musicians, mostly in the field of rock and mostly amateur or semiprofessional in status, to buy Minimoogs. To do this he had to devise de novo the social and technical practices to enable this instrument to be sold. The gift of gab was not enough. As we have seen, his sales practices evolved into a combination of material practices (labeling the instruments), interactional techniques (the hard sell to close the deal), and financial instruments (loan agreements to enable these young musicians to purchase their instruments). To create a market beyond individual sales, he had to establish a dealer network, find a way to instill product loyalty, and teach others how to repeat his success. He had to identify a new group of users and recruit them to take up the instrument. In short, his own claims of inventing the synthesizer market are not that farfetched.

In telling stories about how technologies get developed, we often forget the selling and marketing part. Synthesizers would have remained as tools for elite rock musicians and composers if it had not been for the efforts of Van Koevering, who developed the skills, practices, and expertise to market and sell synthesizers, thus bringing the sounds to a wider audience. As he put it himself, "The sound was in the culture. They heard it on radio, and they heard it on television, not just as the musical sound, but the sound of the synthesizer." Moog is the name we remember partly because Van Koevering made it the name to remember—starting with those silver mailbox letters that the crowds in the clubs saw on the back of the Minimoog.

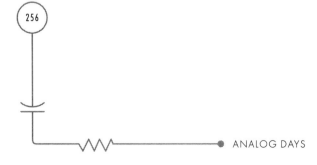

13

Close Encounters with the ARP

It's the only synth that I can operate while I'm drunk.
ROGER POWELL, SPEAKING OF THE ARP 2600

R. A. MOOG CO. and Buchla and Associates were the first companies to sell synthesizers. When retail music stores started selling them in 1971, moving synthesizers from the studios to the streets, other companies entered the fray. In the 1970s the synthesizer industry in both the United States and Europe took off.[1]

ARP was started by an engineer, Alan Robert Pearlman (hence ARP). Pearlman likes to describe himself as a nerd "before the term was invented." He is nine years older than Moog and was born in New York City in 1925. His father was a designer of projectors for movie theaters, and his grandfather (a Russian émigré Jew) made parts for phonographs. Pearlman's boyhood was similar to Moog's; he read *Popular Science* and *Popular Mechanics*, he too was an inveterate tinkerer who built radio sets, using his mother's baking utensils for chasses. He had the obligatory piano lessons and found that his hobbies engaged him more than people did.

The young Pearlman had an insatiable curiosity about how things worked: "I used to haunt the libraries and read what I could . . . whereas other people are interested in social things, nerds are interested in thing things." While Moog was eventually able to overcome his nerdiness to the extent that he liked to work with his customers and was known to hang out with rock stars, Pearlman, from a generation earlier, remained aloof from the world of rock 'n' roll that surrounded him at ARP. He was very much the backroom boy who did all the company's early synthesizer designs. He disliked the razzmatazz of the pop world and preferred to leave the business side of things to others. His musical tastes remained firmly rooted in the classical repertoire.

After a brief spell in the military at the end of World War II, he studied at Worcester Polytechnic Institute—a school with a very strong engineering tradition. During his senior thesis project in 1948 he was able to combine his two loves, music and engineering, for the first time. He designed a vacuum tube "envelope follower" that could extract the envelope of sound from an instrument.

Pearlman put his interest in music aside and took a job for a Boston company working on ionization chambers. As Don Buchla and Hugh Le Caine had found, the world of nuclear physics provided an excellent training ground for the application of electronic skills. Pearlman became involved in analog electronics and started using the newly invented transistors to build devices such as high-voltage power supplies. He audited a course at Harvard University taught by Walter Brattain, one of the inventors of the transistor. He then started building encapsulated operational amplifiers (op amps) using matched pairs of silicon transistors, and shortly thereafter founded his own company, Nexus Research Laboratory, Inc., with another engineer, Roger Noble. Nexus grew throughout the sixties as more and more uses for op amps were found. By 1967 it was grossing over $4 million a year and was sold to a conglomerate. This gave Pearlman the money he needed to start ARP. He used $100,000 of his own money and raised another $100,000 from a small group of investors.

○

A Useful Instrument

All the while, Pearlman had kept up his interest in music: "I played keyboards, I sort of liked them, and I had always been dreaming about how to make different kinds of instruments." He got the idea for ARP after he first heard *Switched-On Bach:* "I said 'Gee that's great!' and I started talking to some people who were in music departments." Among them was Leon Kirschman, a Harvard composer who used the university's Buchla synthesizer. Unimpressed by its lack of a keyboard and failure to stay in tune, Pearlman next talked with another electronic music composer, a former student of Pauline Oliveros's, Gerald Shapiro (Shep). Shep used a Moog in his studio at Brown University. Pearlman: "Yes, it did have a keyboard, but he also said it doesn't stay in tune. 'This Carlos . . . what does he do?' 'Well he tunes up every few measures, but that's alright because it's all taped, spliced.' 'Okay, so wouldn't it be nice if you had a useful instrument that you could take on the stage and would stay in tune?'" Pearlman decided there and then that if he was to make a useful instrument it had to be capable of live performance — that is, it had to stay in tune.

Now that he had a project, he did what all engineers do: he went down to his basement to tinker. He started with oscillator circuits, and soon he realized why Moog's oscillators went out of tune — the key discrete components were not housed close enough together; consequently, temperature variations between different components caused them to drift. By using dual transistors on a single integrated circuit (as he did in the op amp business) Pearlman found he could overcome the temperature gradients and produce a very stable oscillator. This discovery was similar to the one made by Hemsath and Moog as they worked on the Minimoog oscillators.

The greater Boston area was prime recruiting ground for his new company's employees. Also the times really had changed. The success of *S-OB*

259

was generating all sorts of interest in synthesizers among engineers: "We were turning them away." Many of the engineers Pearlman hired were also amateur musicians. Shep continued to play a crucial role—for example, he suggested replacing the patch cords with horizontal and vertical connectors running behind sliders.

○

The ARP 2500

Pearlman now had the two critical innovations of his new instrument: very stable oscillators and a sliding switch matrix system. Pearlman maintains that the array of sliders, all neatly laid out in lines across the front panel, enabled a musician to grasp the overall way the sound was made much faster than with a messy array of patch cords. "It's very much like a graphic equalizer display or something like that." The use of lights and clearly labeled functional pathways etched on the front panel also helped users with little synthesizer experience to quickly understand the instrument.

The ARP 2500 was shown at the May 1970 Audio Engineering Society Convention held in California. Marking the occasion was a full page advertisement in the AES journal that made clear the main selling points of the ARP versus the Moog. "If you would like to spend your time creatively, actively producing new music and sound, rather than fighting your way through a nest of cords, a maze of distracting apparatus, you'll find the ARP uniquely efficient . . . matrix switch interconnection for patching without patch cords." A wicked postscript was appended to the ad: "P.S. The oscillators stay in tune."

The ARP advertisements were as relentless as they were hard-hitting. A new one appeared in the AES journal in October 1970 with an unambiguous message: "If you're a discriminating song writer, recording engineer, film maker, professor, rock artist, composer . . . you don't want notes drifting out of tune. Or messy patch cords hiding the front panel controls. You could use an ARP very soon." The uneasy rivalry between Moog and

Buchla had never produced advertisements like these, implicitly attacking the rival's products. By contrast, a Moog advertisement in the AES journal at the time reveled in the glory of *Switched-On Bach*. The banner headline reads "Long live the Moog!" and has a border repeating the mantra "From Brandenburg to Trumansburg." The Moog advertisement is cluttered, in contrast to the ARP ad's simplicity, perhaps adding to the suggestion that the ARP is the easier synthesizer to operate.

The ARP 2500 turned out not to be in any significant sense cheaper than the modular Moog. Nevertheless, the ARP 2500 system was very elegant and could be expanded in wing cabinets according to customers' needs. Endorsements from musicians ranging from Pete Townshend to Milton Babbitt showed its wide appeal and were conspicuously displayed on the front cover of advertising brochures.

Pearlman's modular synthesizers with temperature-stable oscillators could not have come at a worse time for Moog. Sales of modular systems had already peaked, and here was a new rival threatening to take away what little business there was to be had. Bob Moog had long recognized the problem with oscillator drift and had set in motion a research program to build new oscillators (the 920s). His sales reps, Paul Beaver and Walter Sear, warned him about the high price of his modular equipment and the threat of new competition. But undercapitalized and struggling to survive, Moog could do little to respond.

◯

The ARP 2600

In 1970 Pearlman developed what was to become a classic analog synth. The ARP 2600, a scaled-down version of the 2500, retailed at $2,195. It was a good compromise between the flexibility of a modular synthesizer and the performance capability of a Minimoog. It folded down into a compact luggage-style carrying case and had a built-in amplifier and two monitor speakers.

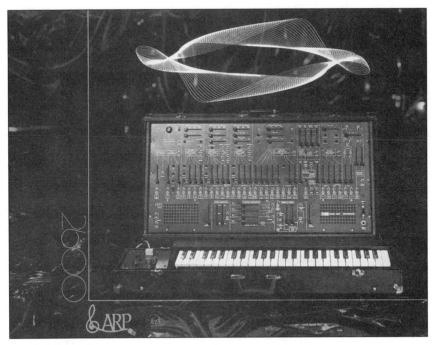

Figure 49. ARP 2600

Pearlman had designed the 2600 with the educational market in mind. He had painstakingly laid out the controls so that the functions were easy to understand. The layout clearly drew on the same inspiration as the Minimoog, with the function of each device, such as VCO-1, VCO-2, VCO-3, the envelope generator, VCF, and VCA all separated off from one another, left to right across the front panel. The use of vertical and horizontal slide pots ensured a precision layout. The simple functional pictures and signal and control pathways drawn on the front panel enabled the structure of the sound to be easily followed. The sockets conveniently placed at all the main inputs and outputs to the modules meant that, if users so wished, they could add patch cords to override the default hard-wired settings. Patch cords combined with hard-wiring enabled Pearlman to draw upon the best features of performance synths *and* the best features of modular synths. Throw in its stable oscillators, compact design, and competitive price (certainly in relation to the modular systems), and one could see that this was a synthesizer to be reckoned with.

The ARP 2600 was introduced at the same 1970 Fall AES as the Minimoog and EMS's VCS3. With ARP's growing success, the Moog company started to get seriously worried. But how had ARP done it? Moog's suspicions turned on the filter, so Rich Walborn, a summer intern, was instructed to take it apart. Jim Scott takes up the story: "They had their filter encapsulated in a plastic block, so you couldn't see it, just a block, no circuit diagrams or nothing . . . so we gave Walborn an ice pick and a whole bunch of methylene chloride and let him pick this thing apart. I think it took him days . . . and, sure enough, they had been deliberately infringing on our filter patent and hiding the whole works inside of a big block of epoxy."

We confronted Alan Pearlman with this story, which has become folklore in the Moog company. He patiently explained that he had also used epoxy for his embedded modules at Nexus, so nothing particularly sinister there. There were technical advantages because it kept the moisture out and the components protected. He added with a half grin, it also "does make it a little bit more difficult to reverse engineer, because we had a lot of competition at Nexus." So what about the filter: had he taken it from Moog? In as many words he admitted that he had. "Well, I came up with a diode version of it on my own. Then I found he was using transistors, so I said Okay, we'll use transistors . . . Uh oh, can't do that. So then we got a sharp letter from a law firm."

Acting on his own patent attorney's advice, Pearlman went back to the Moog Company and had this conversation with Moog: Pearlman said, "'Let's talk about taking out a license.' Moog said, 'Fine, we'll be happy to give you a license,' and he started naming the terms. It was kind of expensive, so then we looked at our own devices and looked at what he was do-

263

ing, we found out that he was doing something that was covered by some of our patents." Parts of the redesigned Minimoog oscillators (particularly the exponential converter) infringed on earlier patents he had taken out on similar devices at Nexus. "So essentially I waved that in front of the lawyers . . . It ended up with a very reasonable settlement between us . . . Not much money ever crossed hands."

Jim Scott maintains that the oscillator circuits on the Minimoog were not in direct violation of Pearlman's patents, but this issue was never tested in court. Scott: "We didn't care about the money, because ARP was so heavy on hype and claiming that they had all this advanced technology that they didn't have, we wanted to force them to put a label on the back that said, 'Filter Circuit Manufactured with a License from the Moog Music Company,' just as a slap in the face. Well, it dragged along so long that by the time it got to the point where we were going to win the case, or had won the case, it was out of production and ARP had changed their filter design back to something else. So it all came to a big, fizzling end."

Edgar Winter was one of the first high-profile rock musicians to use the 2600 (he wore the keyboard around his neck like a guitar). Stevie Wonder played one (with the controls set out in braille), as did Pete Townshend, who used it on several tracks on *Who's Next* (1971); he also owned an ARP 2500. The opening track, "Baba O'Riley," a song named for Townshend's spiritual master, Baba Meher, and for Terry Riley, is one of the best known uses of the ARP in rock. The lengthy little sequenced patch with varying filter that starts the track is immediately recognizable and sets up the contrast for a thundering piano entrance and Roger Daltry's vocals. The song is about alienation, pollution, and a cold, heartless big brother—"only teenage wasteland." It also features a duet between Townshend on ARP 2600 and Dave Arbus (formerly of East of Eden) on violin. Originally, Townshend had intended the song to be a twenty-piece orchestral and synthesizer work, but in the end he went with a demo track he had prerecorded in his own studio on the ARP 2600. As "blue eyes" himself puts it,

"This definitive classic seventies rock song actually came from an indulgent experiment in electronic music."[2]

The ARP 2600 was finding a home not only in rock but in other genres as well. Joe Zawinul of Weather Report, a well-known jazz performer, played two 2600s in counterpoint. Brian Eno, although he never owned an ARP 2600, used it on one track of *Music for Airports* (1978). Eno: "It's a beautiful sound, I think, and one that I couldn't have got from any other synthesizer that I know of."[3]

○

The Odyssey and Pro-Soloist

ARP followed the 2600 with yet another important synthesizer, the Odyssey. This small portable synthesizer with slide pots and two oscillators was ARP's direct answer to the Minimoog. Although it lacked the elusive sound quality of the Minimoog and its rotary pot for pitch-bending was awkward to use, it quickly became a favorite. Eventually the pitch-bend knob was replaced with what became known as the PPC (proportional pitch control) pitch-bending system, which consisted of three pressure-sensitive pads. This more complicated system seems never to have caught on with musicians. The failure of ARP to use pitch wheels is seen by many as a perverse and ill-considered attempt to deliberately *not* do things the Moog way. David Van Koevering, out in the field selling Minimoogs, was not impressed with the Odyssey: "It was a two-oscillator instrument wrapped up in a piece of plastic, and it had a thinner sound . . . didn't have that patented Moog, fat, rich sound, wasn't three oscillators, and it didn't sell well . . . Nobody knew how to sell it." Van Koevering may not be the most objective judge here, of course. Many other musicians have found the Odyssey to be a formidable instrument and, with one of the clearest instruction booklets available at that time, it was many peoples' first introduction to analog performance-quality synthesizers.

ARP went on to introduce an even smaller, cheaper synthesizer with pre-

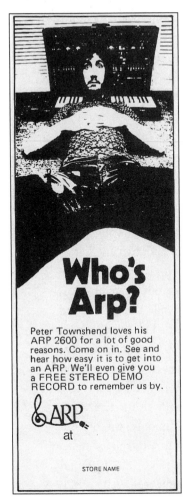

Who's
Arp?

Peter Townshend loves his
ARP 2600 for a lot of good
reasons. Come on in. See and
hear how easy it is to get into
an ARP. We'll even give you
a FREE STEREO DEMO
RECORD to remember us by.

♭ARP.
at

STORE NAME

Figure 50. ARP advertisement featuring
Pete Townshend

set sounds—the Pro-Soloist. This synth was so
small it could just be added as an extra keyboard
on top of an organ. With its built-in keyboard, pre-
set effects, and lack of patching, the Pro-Soloist
was the closest thing yet to being as much monophonic organ as synthesizer. Moog Music responded by producing its own pre-set synthesizer,
the Satellite, which was sold either as a stand-
alone instrument or as an accessory to a Thomas
organ.

○

Dealing in ARPs

ARP now had a complete family of instruments
addressing the needs of users ranging from amateurs to professional recording engineers. The key
to ARP's becoming the dominant manufacturer in
the seventies lay in its marketing. In addition to its
aggressive advertising and its use of musician endorsements, its slick logo of a treble clef with a
power cord running under the letters "ARP" was
also part of the image—and helped establish the
ARP brand name.

ARP knew how to appeal to the new generation
of pop musicians. Alan Pearlman was from a previous generation, but ARP soon hired someone who
could connect with young people. David Friend, a
Princeton graduate student, joined the company
at age 21. He had already worked for RCA labs in New York and had helped
set up an electronic music studio at Yale University. Friend, along with musician Roger Powell, first went on the road for the ARP company to estab-

lish their dealer network. Powell, a studio engineer and talented synthesist, had arrived at ARP one day from Atlanta, Georgia. Powell: "David and I traveled all over the place in a red Chevy van. We tried to sell the 2600s in hi-fi outlets as well as music stores. We got thrown out of most of them . . . The turning point was when [the music retail store] Sam Ash decided to take the 2600, late in 1971. They were really forward looking. They gave us credibility among retailers, and exposed our instruments to all the musicians in New York."[4] Once the 2600 was picked up by rock musicians at New York's most famous music store, the world of rock and the world of synthesizer sales started to reinforce one another with the same synergy that had propelled the Minimoog into the stores.

ARP was also the first synthesizer manufacturer to realize the importance of and capitalize upon National Association of Music Merchant shows. While Bob Moog in 1971 felt like "a fish out of water" at his first NAMM convention, ARP seemed quite comfortable swimming around in this new pond. Bob was impressed by their slick demonstration: "ARP was down the aisle from us and they were really hip because they had actually worked out a demonstration. You know, they *played music* on that—cornball, everyday, pop music." Alan Pearlman loved the NAMM shows. Some of his fondest memories of the synthesizer business were the impromptu NAMM jams, where his musicians would get together to play after the day's business was over.

By 1974 ARP had set up an impressive distribution network in the United

Stevie Wonder loves his ARP 2600. It's got a four-octave keyboard that lets him bend notes and really wail. His ARP helps him create any sound he's ever dreamed of, and 1001 yet to come. And he plays his ARP knowing it will never go out of tune. No wonder Stevie ARPS. Now how about you? Come on in. Let us tell you all about the new music . . . plus show you how easy an ARP is to get into. We'll even give you A FREE STEREO DEMO RECORD to remember us by.

World's leading manufacturer of electronic musical synthesizers. That's ARP.

it up at

STORE NAME

Figure 51. ARP advertisements featuring Stevie Wonder

States and worldwide and was describing itself as the "World's Leading Manufacturer of Electronic Music Synthesizers." ARP supplied its dealers with special promotional kits giving precise instructions on how best to demo the instruments. Special ARP promotional records were issued and service centers and finance plans put in place. Dealers were updated with the successes of the company via a newsletter, ARPEGGIO. This publication featured unremitting self-promotion for the company, with prominently displayed pictures of musicians such as Jimmy Page, Pete Townshend, Edgar Winter, and Stevie Wonder with their ARPs. It ran one item proclaiming that ARP "is almost synonymous with synthesizers" and, as evidence, listed famous musicians in England, Germany, Israel, and Japan who had bought ARPs. ARPs were even selling in Iran.

Was the ARP name now synonymous with the synthesizer? Not quite. Bob Moog recalls a story in a newspaper at the time that symbolized the problem ARP faced: "I can remember a big newspaper spread about Herb Pilhofer. On the left-hand page it said, 'The Composer,' with a picture of Pilhofer. And on the right-hand page it said. 'And his Moog,' with a big picture of an ARP system. I used to get a kick out of that."[5]

As well as focusing on pop and amateur users, ARP continued to develop the educational market. Moog and Deutsch had also tried to design a synthesizer for the education market (the so-called Ed. Moog). The target market for ARP was schools with small or medium-sized music departments—those who could afford to spend $1,000 on an Odyssey. ARP produced special brochures and workshops aimed at schools. The *ARP Educator's Brochure* listed ten different ways a synthesizer could be incorporated into classroom activities, including psychology classes, where it was suggested that "the effect of different sounds upon laboratory animals offers exciting contexts for experimentation and discovery." The lab rats must have loved their new ARPs.

In 1973 Moog Music had been taken over by the musical instrument giant Norlin, which treated the synthesizer as just another musical instrument. With Van Koevering having departed (he left when Norlin took over), no one seemed to know how to sell synthesizers any more. Scott: "There was a serious problem . . . Norlin's other businesses were Pearl drums, Gibson guitars, Armstrong flutes, where the music dealer bought them at price A, then you sold them at price B—a profit! And synthesizers couldn't be sold that way, they required a lot of education, both of the dealer and the market . . . you hype it in the local press and have a bunch of people show up and then you put on a big presentation, get everybody all fired up." The net effect was, according to Jim Scott, that "ARP, with a product that didn't sound as good, wasn't as reliable, was more expensive, beat our pants off because they were doing the marketing right."

While anyone, at any time, could walk into Moog's funky Trumansburg factory, ARP was run much like any other business. Borden: "Much more urban, hipper. You walked in there and it was, like, wall-to-wall carpeting with people in suits . . . You know, and also they were kind of condescending to the whole Moog thing, you know."

And all this activity was working for ARP. Its sales increased from $1.5 million in 1972 to $2.5 million in 1973 to $3.5 million in 1974. At this rate of growth, the company would indeed become, as it boasted, the biggest single synthesizer manufacturer in the world. It employed 83 people and had moved into a brand new 50,000 square foot factory space. Rock stars regularly visited ARP headquarters and, according to most reports, the owners themselves reveled in the good times. The list of well-known musicians who used ARP synthesizers was truly extraordinary and included names from many different genres of popular music.

◎

Tom Oberheim, the Accessories Industry, and Polyphony

By the mid-seventies a secondary market in devices that could be added to synthesizers developed. This accessories market was started by an ARP dealer, Tom Oberheim, who later became famous in the synthesizer industry as the first inventor of a truly polyphonic synthesizer. Oberheim was a Los Angeles-based computer engineer who started to build ring modulators for musician friends and formed his own company after his ring modulator was used in the movie *Beneath the Planet of the Apes* (1970). Oberheim traces his involvement in synthesizers to the influence of Paul Beaver, who helped him build a digital phase shifter that also emulated the effect of Leslie speakers (rotating speakers first used with Hammond organs). His Maestro phase shifter became a huge success—he sold 25,000 within three years.

At the 1971 NAMM show Oberheim approached ARP and "almost as a lark" asked to be the company's LA dealer. "I actually became a one-man ARP dealer for a couple of years, and in doing so I learned the ARP 2600 very well." Oberheim was soon successfully showing and selling the ARP to his music contacts in LA: "I started selling ARP 2600s to essentially the same people I'd met selling my ring modulators, I'd call them and say, 'Hey, I've got this synthesizer.' 'What's a synthesizer?' And I'd stick it in the back of my car and I'd drive it over to their place and I'd show it to them and then they'd buy it."

Oberheim was fascinated by his ARP 2600: "I took it home one night, put it in my bedroom, and let it run on sample and hold all night."[6] As an ARP dealer, he had access to the ARP's schematics, and he noticed that the ARP 2500 had a little feature which allowed two pitches to be played at the same time. He designed a kit to add this feature to the ARP 2600. This little kit modification was to have an important effect on the future of the synthesizer. At the time Oberheim was doing lunchtime concerts with a friend from UCLA, keyboardist Richard Grayson. In the first half of the concert Grayson took a theme from the audience and then reworked it in the style of some classical composer: "You know *A Hard Days Night* in the style of Mozart or whatever." In the second half Grayson and Oberheim improvised using ring modulators and a couple of tape recorders doing tape delay: "It was pretty wild."

One day Oberheim introduced his newly modified ARP 2600; Grayson played two of them stacked one above the other (each capable of two pitches): "We did a sort of poor man's version of kind of a *Switched-On Bach* thing, and it was like extremely crude, but it was great . . . at that point in time there were no commercially available polyphonic synthesizers." This set Oberheim thinking. "And one of the things that was always kind of going through my mind is, I had all this digital computer background I had no way of using. So it occurred to me at some point to build a digital sequencer that would hook up to the ARP 2600. And so I . . . designed this digital sequencer."

His DS-2 sequencer eventually had the capability for storing 144 notes. Oberheim quickly discovered that in live performance the sequencer would take over the synthesizer, using up the one voice and leaving the performer nothing to play over the sequencer. This led him to develop a small inexpensive little synthesizer, the Oberheim "Synthesizer Expander Module" which could be added to a Minimoog or ARP along with his sequencer. In 1975 Oberheim "expanded" his expander modules into his "Two Voice" and "Four Voice" instruments, the first polyphonic synthesizers to use individual voltage-controlled circuits for each voice. Oberheim (in partnership with Emu, which had developed a polyphonic keyboard) went on to introduce a range of polyphonic synthesizers, including the OB-X (1979), one of the first completely programmable polyphonic synthesizers.

○

ARP's Fall

Alan Pearlman loved designing synthesizers and was happy to leave the day-to-day running of his company to others. This, in the long term, turned out to be a huge mistake. The demise of ARP can be traced to a power struggle between three men, each of whom thought he was running ARP. The three were Pearlman, David Friend (who had become president), and

Lewis G. Pollock, legal counsel whom Pearlman had first used to negotiate the buy-out of his previous company, Nexus. Pearlman was disconcerted to find that Pollock was spending more and more time on ARP business and that he had maneuvered himself into a leading executive position.

ARP's downfall can be traced to the ill-fated Avatar project (which Alan Pearlman opposed) to develop a guitar synthesizer—in other words a synthesizer that used guitar strings as the controller.[7] This was not such an unreasonable idea on the face of it, given that there were many more rock guitarists than keyboardists. But the company vastly underestimated the technical difficulties and sunk more and more money into the project before going under in 1981.[8] Lack of capital was also endemic. In the end ARP was forced to declare bankruptcy, and Alan Pearlman, his friends, and family lost $500,000 in cash. Creditors and stockholders lost nearly $4 million. Pearlman to this day is still paying off losses from his adventures in the field of synthesizers.

But he is not bitter. He told us that he enjoyed the first five years of his business, but when company in-fighting raised its head, most of the joy left. Also, he was bewildered by the uptake of his synthesizers in the world of rock. Sounding like everyone's dad in the sixties, he complained to us about not being able to hear the words of rock because of the decibels. Like Moog, he too had had private hopes that the synthesizer would turn into a "serious instrument" for classical music. The success of his 2600 was thus a double-edged sword.

No doubt lessons in how not to run a business can be drawn from the demise of ARP, especially regarding control of R&D (research and development), but the bigger lesson of ARP is that for years they ran an extremely successful synthesizer business. They simply extended what Moog had already set in motion. ARP, of course, had an advantage as a latecomer to the synthesizer industry: it could learn (and pilfer!) from what Moog and Buchla had done before. ARP entered just at the right time, when synthesizers were truly on the road to becoming a mass-market item. Once you had the product, establishing dealers and marketing were the keys to the business, and ARP used them to open a lot of doors. The company worked closely with musicians and listened to what they wanted; it employed engineers who were themselves musicians; it carried out engineering innovations; it produced a reliable, well-engineered product supported by sales and service; and it tried to place this product into as many markets as possible and en route developed new markets, such as amateur home users and the educational market. ARP was the first synthesizer company to realize the importance of going to NAMM and having a slick demonstration that appealed to retailers. The company had global ambitions and helped bring the synthesizer to the world. And like Moog, ARP recognized that the synthesizer was, in part, an instrument of youth rebellion.

○

Space Soundscapes: Close Encounters and Star Wars

One of ARP's publicity coups was the appearance of an ARP 2500 (and ARP's chief service technician, Phil Dodds) in Steven Spielberg's sci-fi blockbuster *Close Encounters of the Third Kind* (1977). The ARP logo can be seen clearly on the giant synthesizer whose five note sequence was used to make contact between aliens and humans. It further reinforced the association between the synthesizer and space and got ARP some massive free publicity.

The ARP 2600 was used for the sound effects and voice of the lovable little robot R2D2 (Artoo) in one of most influential sci-fi movies ever, *Star Wars* (1977). George Lucas regarded the sound effects to be so important that he hired Ben Burtt to make them a full two years before the movie was due to be released. Burtt was new to the synthesizer and borrowed an ARP 2600 from Francis Ford Coppola for the duration of the movie. Burtt found the envelope generator to be particularly important in ensuring consistency in the strings of sounds that made up Artoo's voice. Burtt: "The voices of the robots, most notably Artoo. They were all derived from the ARP, from combinations of organic sounds and ARP sounds, sometimes played simultaneously . . . The important thing was to get comprehensible emotions . . . Whenever you hear him a connection has to be made, like, 'Oh I understand. He's excited,' or, 'He's mad,' or 'He's laughing.'"[9]

Burtt made a lot of other sounds for *Star Wars* on the ARP, including the "sounds of computers" and the ambient sound in the spaceships, in the cockpit, and in the Death Star power plant. Sometimes he combined naturally found sounds with electronically generated sounds. For instance, for the spaceship pass-bys, "I used my favorite sound source the Goodyear blimp. It's got this low groan and two engines that beat against each other. It's a terrific sound. I slowed it down and flanged it to get the Doppler effect as it went across the screen." Spaceships, of course, don't make any sound at all in a vacuum, but the pass-bys heard on *Star Wars* have now become the de facto standard for what a credible spaceship should sound like. One of the interesting points made by Burtt is that sounds must sound credible—he saw a version of *Star Wars* without the sound effects and found that the movie became a comedy.

The process of making sounds for movies by electronic means was first developed by Louis and Bebe Barron in 1956 when they used cybernetic devices of their own design to make the score for *Forbidden Planet*. It was taken to a new level with the arrival of the synthesizer. So many more sounds were feasible. But there was a problem with this multitude of sounds, a problem similar to that which studio musicians faced in the early days of the Moog. Beaver and Krause, for example, could produce a huge array of sounds in the studio, but the producers' and musicians' vocabulary for recognizing and describing those sounds was extremely limited. The vocabulary to go with the new synthesizer soundscapes had not yet been devised. The visual dimensions of films and TV provide a way for some of this soundscape to become reproducible, that is, recognized, described, and passed on. The success of *Star Wars* meant that a new part of our soundscape was created. A new vocabulary for recognizing and describing space sounds developed that other synthesists could use.

For instance, Suzanne Ciani reports, "Everything is defined by being successful. I remember before the *Star Wars* thing, before there was specific vocabulary defined for all the sounds that *Star Wars* made popular, it was hard for me to tell people that the sound I had was the sound of a spaceship going by. Somebody would always ask what it was. Now it's part of the sounds that people are familiar with. Now I go to sessions and people ask for Artoo Detoo sounds."[10]

The new soundscapes were reinforced with visual images (movies and television) and acquired their own new terms (like "spaceship sound") which many synthesists could now produce on demand with any synthesizer (in Ciani's case, the Buchla 200). The use of synthesized sounds is today commonplace in movies and television. Shows like *The X-Files* and *Who Wants to Be a Millionaire* are replete with synthesized sounds to add mood, atmosphere, and excitement. These programs serve further to define the vocabulary of soundscapes, or the "poetry of sound" that pioneers like Ciani helped evoke in another industry. All this was the legacy of the suitcase-sized machine.

14

From Daleks to the Dark Side of the Moon

All the rain pours down amen on the works of last year's man.
LEONARD COHEN

ELECTRONIC MUSIC STUDIOS (EMS) of Putney, London, was a peculiarly British operation. Its founder, Peter Zinovieff, was the son of a Russian aristocrat who had escaped the revolution to settle in London. At the beginning of the sixties Peter had the financial means to do whatsoever he desired. And what he desired more than anything was to advance the field of electronic music. He had one of the most powerful computers to be found in private hands and one of the best private electronic music studios anywhere in the world. Working for him was an unassuming engineer, David Cockerell, who designed EMS's most famous synthesizer, the VCS3 (Voltage-Controlled Studio, attempt #3). As usual, a musician was also in the picture—Tristram Cary, a composer and teacher at London's Royal College of Music.

The Putney studio where EMS had its headquarters was unique because the line of synthesizers it produced was designed and manufactured elsewhere. The studio was run more like a salon than a business. It was housed in the same pair of Victorian terraces where Peter, his wife Victoria, and their three children lived. Electronic equipment, antique furniture, and fine wine helped create the unique ambience of EMS's famous lunches, where rock stars mixed with leading composers. A more different environment from Bob Moog's funky factory or ARP's slick American corporate offices one could not imagine.

○

David and the Daleks

David Cockerell, born 1942, had the perfect career trajectory for a synthesizer designer: built crystal radios in his bedroom as a boy, tried to learn the violin and piano, played bass in a rock 'n' roll group (The Hot Cocoas), for whom he also built amplifiers, and dropped out of his college course in philosophy to take a day job as a technician with the Ministry of Health so that he could attend night school to study electrical engineering. He discovered that his real talent was building electronic devices for making

Figure 52. Daleks in London; Tom Baker (inset) as Dr. Who at the controls of the Tardis

music. After starting off with a little electronic music box that made random sounds, he moved on to tape-delay reverbs and then ring modulators. Why ring modulators? "I wanted to make the sound of a Dalek."

Long before there was Darth Vader there were Daleks. These half-machine, half-alien creatures—squid-like things encased in cyborg armor—used to terrify British school kids brought up in the early sixties on the weekly TV show *Dr. Who*. The Daleks had megalomaniacal ambitions. Imagine R2D2 gone bad, really bad—a mobile, over-sized traffic cone with weird protuberances dispensing lethal rays. As a Dalek swung round to blast some hapless earthling, it would scream in an excited mechanized voice, "EX-TER-MIN-ATE, EX-TER-MIN-ATE." At this point the black and white contrast of the TV set would reverse and death would surely result. Believe it or not, the Daleks were very, very scary, and every British school kid wanted to imitate that voice.[1]

○

Peter Gets an Education

Britain is a society built on class, and education is usually the crucial marker. If you went to Gordonstoun School in the North of Scotland, as Peter Zinovieff did, you were upper-class. Gordonstoun is the school attended by Prince Charles and his father, the Duke of Edinburgh. In Britain, "public schools" (that is to say, fee-paying schools) like Gordonstoun do not pamper their clientele; they are rugged environments. The English upper classes believe that the tougher the schooling, the harsher the conditions, and the more money they are required to pay, the better the education must be. Cold showers, corporal punishment, and rugby fields are meant to breed the sort of character needed to run an empire—even if there is no empire left to run.

Peter's parents had arrived in Britain to escape the Russian Revolution. His mother, a princess, had met Peter's father, another aristocrat, in London; they got married, gave birth to Peter in 1933, and then divorced. Young Peter was left in the hands of grandparents and then sent to a Dorset convent school to escape the bombs falling on London during World War II.

Peter tells horror stories about that school. Separated from his family, he found life at the convent hard; he was locked in a cupboard and beaten every day by the nuns. Many years later he got some sort of revenge on his childhood tormentors. One of the advertisements for the synthi range of synthesizers that EMS produced showed a picture of a nun playing a synthesizer with the caption, "Every Nun Needs a Synthi." Peter Zinovieff never lost his sense of humor.

After the war he lived with his father on a farm in Sussex and attended Guilford Grammar School (a state school for bright kids). There, Peter explored his hobby of amateur radio, starting by putting together crystal sets and ending up by building a transmitter. "And that smell of that old-fashioned solder and those boards, and the big thick wires, and the big transformers for high tension . . . it was lovely." His pleasant life in Sussex was soon to end. Guilford Grammar School "wasn't considered posh enough by my posh grandmother, stepmother rather, and I was sent to a really tough school, Gordonstoun." Gordonstoun may have been a challenge, but it was not as harsh as the convent had been. He ended up loving Gordonstoun for its sailing and climbing.

Peter went on to Oxford University, where he became the "the eternal student." He ended up with a PhD in geology. Having learned to play piano as a child, at Oxford he got interested in experimental music, forming his own little group, Biscuit Tin (the name comes from the biscuit tins they hammered on and shook around). He played prepared piano and used his Grundig tape recorder to record the music they made (he would speed it up and slow it down but never spliced).

Peter eventually married "a very beautiful girl, Victoria, who happened to be immensely rich." Victoria Ross was only 17 when they got married, and her wealth shaped Peter's early career. He was offered teaching and research posts overseas, but Victoria's parents wanted their daughter to stay in England. He ended up working as a mathematician for the Air Ministry in London. He found the work fascinating but "nasty"—his job was programming for various nuclear war scenarios. Although Peter claims not to be a very good mathematician, he did acquire a deep interest in the mathematical properties of random numbers. The war games lasted only a year; given Victoria's wealth, it made no sense for him to continue in paid work, since all of his income was being swallowed up by their joint taxes.

○

Peter Gets Passionate

Peter was now in the fortunate position of being a kept man—he could do whatsoever he wanted, as long as it kept him in England. So he decided to return to his Oxford hobby and make electronic music. "I got a couple of tape recorders and started recording piano sounds, using microphones. And I got very quickly passionate, really passionate." It was 1960 and London was awash with surplus equipment ideal for Peter's hobby: "In those days there was a wonderful street, Lisle Street, which is now part of China-town London, and it was just full from one end to the other of second-hand, ex-Army, electronic stores . . . And I got amazing things, sound generators, noise generators, fantastic wave analysis machines . . . great big, huge bulky things."

London was the home of the BBC Radiophonic Workshop, where the special effects for BBC programs were made and where the Dalek voices were first created. The signature tune for *Dr. Who*, realized by Delia Derbyshire, became one of their best-known theme tunes and ensured that nearly all British kids had exposure to at least one unforgettable piece of electronic music.[2] One of the pioneers at the BBC Radiophonic workshop was Daphne Oram, and she taught Peter how to make electronic music.

Like Don Buchla, he was struck by how inefficient the whole splicing process was. He even put together a crude form of sequencer using old analog telephone exchange equipment. But his own electronic skills were limited, and he was dissatisfied with this electromechanical device, which worked badly. He soon found a technician, Mark Dowson, whom he paid to make an electronic version. Mark went on to make a random number generator for Peter that ran from the radioactive decay of a luminous watch.

Dowson turned out to be a crucial contact because he was a childhood friend of David Cockerell: "Yeah we'd discuss volts and amps when we were knee high." Soon David was building devices for Peter: "He was a complete genius. You could explain a problem, and in a few weeks the problem would be solved, even if it sounded very extravagant." One could hardly imagine a better combination, the articulate dreamer with a seemingly bottomless purse and the down-to-earth technician who could turn others' electronic dreams into reality. The former Hot Cocoa and Biscuit Tin players were clearly destined for each other.

◎

David and Peter

Two crucial elements of Peter's philosophy for making electronic music were now in place, sequencing and randomness. The third element was sampling. Peter wanted to make a sampler because he shared the view of the *musique concrète* practitioners that "real sounds have got so much complexity that they're better than synthetic sounds." His idea was to sample sounds and then analyze and resynthesize them. This project would obsess Peter for the duration of EMS's history, and all other EMS projects were a means for Peter to work toward this goal. He got David and his successor at EMS, Peter Eassty, to build a range of different analyzers, including eventually a digital analyzer: "It worked for two minutes maximum, and that was the end of its life."

In order to build a digital analyzer, they first needed a computer. They bought a DEC minicomputer, the PDP-8, which had just come out. Purchasing this computer stretched even Peter's seemingly unlimited resources. Once more Victoria came to the rescue: "She had her own beautiful pearl and turquoise tiara and we sold that for the computer." Peter eventually bought another DEC computer, and the two computers acquired names, Sofka and Leo—the same names as two of the couple's three children.

The plan was for the computer to drive a bank of filters and oscillators—filters to analyze the sound and oscillators to resynthesize it. Peter got David to build a bank of 64 filters, which in the resonant mode could also be used as sine wave generators. It is here that David drew upon what Moog had done earlier. He was a keen follower of American developments in electronics: "I saw an article by Bob Moog in *Electronics World* . . . and [it] revealed the principle of exponential control, which I'd never thought of before."

◎

Unit Delta Plus

Peter kept up his connections with the BBC Radiophonic Workshop, eventually teaming up with two of its staff, Delia Derbyshire and Brian Hodgson, to form their own little company, Unit Delta Plus. The idea was to pool their equipment and make commercial jingles. They made one for Phillips, which was used to advertise electrical appliances on TV, but Peter found he was not commercially inclined. Unit Delta Plus did, however, have some success performing their own compositions.[3] They once shared a bill with Paul McCartney at a "Million Volt Light and Sound RAVE—Dancing to Mystic Rock Groups" held in early 1967 at one of London's most hip venues, the Roundhouse. At the time, McCartney was deeply involved with electronic music (Stockhausen is one of the figures on the cover of *Sergeant Pepper*) and was a regular visitor to Peter's studio.

As the Swinging Sixties started to unfold around Peter, he displayed an almost studied indifference to the rock musicians who became interested in electronic music. The world of rock stars, drugs, and counterculture was not really Peter's world. He was bemused by the EMS secretaries swooning

over the famous musicians who came to visit, but for Peter it was business as usual.

○

Putney

In 1968 David Cockerell finally got his electrical engineering degree. He now started to work full time designing devices for Peter from his Cricklewood home/workshop. Peter, by now, had moved the studio from his house in Belgravia to its famous location at 49 Deodar Road, Putney, south of the river Thames. Part of the first studio Peter built actually overhung the Thames, so at high tide he could see its murky waters flowing by below.

Peter worked with many composers, including Hans Werner Henze and Harrison Birtwistle. Stockhausen was a frequent visitor, and Peter was amused to recall that Stockhausen "arrogantly thought he could work all the machinery . . . of course, he couldn't." Peter's collaboration with Birtwistle, a close friend, led to "Chronometer" (1971), a piece based on recordings of Big Ben that Peter analyzed and resynthesized using his own MUSYS programming language.

Peter regarded his studio as a cutting-edge research institute. But maintaining such a studio was expensive. As Peter's projects sucked in more and more of Victoria's funds, at some point she put her foot down. Peter had to find new ways to support the studio. That was where the VCS3 came in.

○

From Junk to the VCS3

One composer attracted to Peter's studio was Tristram Cary, who also was a former student of Daphne Oram's. The son of Irish novelist Joyce Cary, he had a wealth of experience making commercial and avant-garde electronic music (he did the film score for *The Lady Killers*, 1955, and *Quartermass*

283

and the Pit, 1967) and owned a private studio. Tristram initially worked as a consultant for Peter and was there one night in the pub with David and Peter when the idea for what became the VCS3 was dreamed up.

Cockerell remembers building the VCS1, or Voltage-Controlled Studio Mark One. "We made one little box for the Australian composer Don Banks, which we called VCS1 . . . And we made maybe two or three of those . . . It was a thing the size of a shoebox with a lot of knobs, oscillators, a filter, not voltage controlled, maybe a ring modulator, and envelope modulator." No one can remember if there was a VCS2, but if there was it was probably an expanded VCS1. Peter thinks it's quite likely that they moved to the VCS3 simply because they liked the sound of the name (which was invented by Tristram): "it just sounded better than VCS1."

Much of the original idea for what should be on the VCS3 was hammered out in the pub. The idea was to build a synthesizer that could be sold to schools. It would have a simple layout with three oscillators, a filter, an envelope generator, and a ring modulator. Tristram was convinced that a market for a machine like this, for teaching purposes, could be found—the same market that Moog had first noticed and that ARP was starting to reach in the United States. To be bought by teachers, however, the machine would have to be cheap. The Moogs and ARPs (which were even more highly priced in the UK than they were in the States) were out of the price range of most UK schools.

Peter remembers that he specified the functions that were needed, Tristram designed the case, and David did all the electronics. David remembers that the goal was to make the VCS3 as inexpensive as possible: "A lot of the design was dictated by really silly things, like what surplus stuff I could buy on Lisle street . . . For instance, those slow-motion dials for the oscillator. That was bought on Lisle Street, in fact nearly all the components were bought on Lisle Street."

It's an extraordinary fact, but like the first version of America's most popular portable synthesizer, the Minimoog, the prototype of the VCS3,

Britain's best known portable synth, was also first assembled out of junk. Here the role of the hobbyist tradition from which engineers like David Cockerell had emerged was crucial. As he told us, "Being an impoverished amateur, I was always conscious of making things cheap."

As with the development of the Minimoog, chance played its part. The most distinctive feature of the VCS3 is its 16 × 16 matrix panel, where everything is connected by little pins, thus avoiding messy patch cords. David: "It was serendipity. I thought that would be a good way of connecting inputs to outputs and much more convenient than a set of cords." The first matrix panel was found in a surplus store on Lisle Street.

David had already designed a ring modulator and a voltage-controlled oscillator. "I worked on them [the oscillators] a lot to make them more accurate, although I didn't entirely succeed. In fact they were awful . . . [but] if you're not really interested in pitched music it doesn't matter." The filter was based on the ladder filter design that Moog had published in *Electronics World*. David: "I saw the way Moog did it, but I adapted that and changed that . . . he had a ladder using ground-base transistors, and I changed it to use simple diodes . . . [to make it] cheaper. Transistors were twenty pence and diodes were tuppence!"

Another notable feature of the VCS3 is the little X-Y joy stick that sits in the right-hand section of the base panel. Again, David's hobbyist background helped out: he simply adapted a joy stick from a radio-controlled model airplane. He thought it was a good way of controlling two parameters and soon persuaded Peter of its merits.

Tristram designed the case: "Somewhere along the line, probably on a scrap of paper in a pub, the miniature desk shape seemed a good idea, and having worked out suitable dimensions with David, I took home a patch matrix and spent a weekend at Fressingfield [his home studio] making the box for the prototype."[4]

The VCS3's three oscillators produce a combination of sawtooth, sine wave, and square wave outputs. It has a built-in amplifier, speakers, and

285

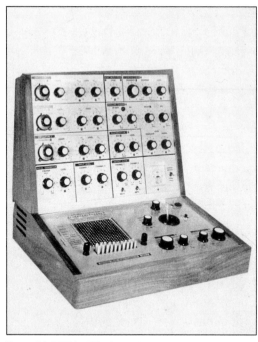

Figure 53. VCS3—"The Putney"

a way of panning between the two stereo speakers. Another feature that turned out to be important was the audio stereo inputs for processing other sources of sound. The layout is very easy for the user to understand. Each module is bracketed by lines etched on the front panel; nothing is cluttered. This layout, with the three oscillators aligned down the left-hand side of the panel, was developed earlier than Bill Hemsath's design for the Minimoog. An excellent manual, which Tristam Cary wrote, accompanied each instrument. The desk console shape of the VCS3, with the joy stick standing out, evokes the controls of a spaceship. Its knobs, set in aluminum slow-motion dials, invite someone to twiddle them. The VCS3 was released in November 1969 around the same time EMS as a company was formed.

Five VCS3s were handwired before David finally laid out the circuit boards. The woodwork and etching of the panels were farmed out to small companies around London. One company in Wareham (150 miles from London), Hilton Electronics, run by Robin Hilton, built transformers for the VCS3s and eventually took over all manufacturing. The production runs were initially small, David remembers, "maybe a dozen a month, that sort of thing." The goal was to sell the instruments at around 200 pounds each; eventually they retailed for 330 pounds. Still, this was much less expensive than ARPs and Minimoogs. The Synthi A (originally named the Portabella) was a fully portable VCS3, built into a briefcase and sold retail at 198 pounds.

Going Backward

The first VCS3s did not have a keyboard. As David remembers, "The modern avant-garde composers weren't much interested in keyboards or well-tempered music." The DK2 touch-sensitive keyboard was added later. It had a variety of tunings and could even be set up in a reverse mode so that the harder you hit the keys the softer the note sounded. This keyboard, which is the first commercially available touch-sensitive keyboard built for a synthesizer, was not easy to use.[5] It drew a sharp response from no less a person than Walter Carlos who, in a damning review of the VCS3 for the *Whole Earth Catalogue*, did not seem to realize that the keyboard was deliberately designed with the option of dynamic reversal. Carlos wrote: "The PUTNEY is a real toy. Its components are highly unstable/unpredictable, and the selection made is highly gimmick oriented . . . It also has a so-called touch-sensitive keyboard which has to be tried to be believed, it's that awful! No feel or physical feedback at all (as there is on a piano, for example); again, another great concept worked out in ignorance (and the one I tried worked backwards: softer touch = louder sounds!) But it *is* small & portable & groups might like it for special effects."[6]

The VCS3 never won widespread adoption as a keyboard instrument, and keyboard music did not interest the composers behind EMS. The story David tells about the first time *Switched-On Bach* was heard in London summarizes the gulf between the different genres of music for which the synthesizer could be used. The occasion was a visit to London by Bob Moog. "I remember once Ray Dolby, Harrison Birtwistle, and others came to Peter's house and then we went to Ray Dolby's factory and Moog gave a little illustrated demonstration, for which he played *Switched-On Bach*. And Harrison Birtwistle was outraged, and stormed out of the room, slamming the door behind him." A serious composer, Birtwistle was unimpressed by this use of the synthesizer. Cockerell's reaction, on the other hand, was very different: "I thought it was great."

The VCS3 had a rather odd relationship to its creators. Peter told us he never used it for compositional purposes since his own studio equipment was well in advance of anything the VCS3 offered. For David, too, the VCS3 was no technical advance: "I'd been making more interesting and intricate machines for Peter for his studio for some time. This VCS3 was really going backwards a little bit." We asked David if he liked the sound of the synthesizer he had built. "I liked the weird, spacey sounds, although they're not what I describe as music."

◎

The Synthi 100

Other products followed. In 1971 EMS produced the largest commercial modular analog synthesizer the world had yet seen—the Synthi 100. Basically "a VCS3 times 10," it sold originally for 5,500 pounds. David remembers that he did improve on the modules quite a bit, using newly available integrated circuits. Don Buchla remembers Peter Zinovieff coming to the United States with a Synthi 100, an instrument that Don Buchla thought was "the most insane thing" to try to sell in the States. As he remembers, there were no takers, and Peter was forced to give it away in the end to avoid paying the shipping charges back to the UK.

The Synthi 100 did, however, lead to other products, including EMS's 256-step digital sequencer—the first such instrument to become commercially available. When this was added to the Synthi A along with a touch-plate keyboard, the Synthi AKS was born.[7]

◎

A Family of Synthis

Like ARP, EMS now had a family of synthesizers; furthermore, it had a family name for them, the Synthis (Tristram's idea). It was Peter who came up with the "Everybody Needs a Synthi" series of advertisements, which ran in British sound and music magazines. Each wave of ads showed a dif-

ferent group of people with their synthe-
sizers. It included Every Band, Every
Group, Every Picnic, Every Nun, Every
Dream, Every Concert, Every Conduc-
tor (drawn by Harrison Birtwistle), Every
Opera, Every School, Every Orpheus,
Everybody, and Every Note. "Every
Christmas" was added as the company
Christmas card.

Peter himself thought that the "Every
Note" slogan was inspirational, and he
had EMS print it on all their pencils.
The idea was to get away from the synthesizer as some high-tech device and
show it as part of the normal pastoral life
of England. The striking contrast between the high-tech synthesizers and
the settings no doubt contributed to
making the advertisements memorable.
This low-key whimsical, almost child-

Figure 54. Early advertisement for EMS synthesizers

like way of drawing attention to their products was no doubt in tune with
the sentiments of the British hippy rock musicians, academics, teachers,
and avant-garde artists drawn to EMS's synthesizers. British psychedelia itself was much more whimsical, pastoral, and full of references to childhood than the U.S. movement. This was a very different way of selling synthesizers than that followed in the States by Moog and ARP.

◎

Electronic Music in Britain

Tristram and Peter also saw the importance of nurturing the wider field of
electronic music in Britain. At Queen Elizabeth Hall in 1968 they staged

the very first live electronic music concert ever held in London. The star was Peter's computer, which just sat on the stage alone, playing "Partita for Unaccompanied Computer." "We wondered if we could just get 200 people to pay for it. But we had to turn away 300 people or so, and it was full, it was 1,100 people." Building on this success, they put on a whole series of concerts at the same venue, each wilder than the next. Peter, like Ramon Sender in San Francisco, discovered that "anybody who has listened to electronic music a lot knows that it's nearly 100 percent terribly boring. And so whatever tricks you can do to make it less boring and give people ice creams to eat and they'll like it more." Some of the tricks he recalls included a concert where the programs were made out of shiny but brittle silver paper on the outside, "so when anybody looked at it it made a crinkling noise, like 1,100 people crinkling, it was really wonderful, and there were sort of stochastic poems where people would have to read out bits . . . and it was a lot of audience participation . . . One we had a four-poster bed and two or three beautiful girls making love in it . . . you know humping around under beautiful covers."

These concerts, featuring leading composers and audience participation, are reminiscent of events that the San Francisco Tape Music Center was putting on at about the same time. Peter recalls that the audience was a mix of hippies and curiosity-seekers. Although the London counterculture at places like the UFO and Roundhouse was developing all around EMS, Peter's studio was largely remote from such activities and did not play the decisive role in shaping events that the Tape Center did.

○

The Special Putney Ambience

Unlike Moog and ARP, EMS had its studio at the heart of a capital city. Peter found that apart from the advertisements, he did not have to do much in the area of marketing. Certainly recruiting a sales force was not his modus operandi. EMS had their synthesizers on display in Macari's Music Store in

Figure 55. EMS studio, Putney, London

Charing Cross Road—at that time, if you wanted to buy a synthesizer you would have to come into London. It seems never to have occurred to them to market through other retail music stores. They did send a direct-mail brochure to schools, with a letter of introduction from Tristram. Ninety percent of all sales of the VCS3 were to schools.

It fell to Robin Wood, a university dropout who joined the company in the role of an odd-jobs man, to demonstrate EMS instruments at the studio. Peter and Robin evolved a kind of double-act, with Peter doing the talking and Robin the demonstrations. Once word spread, anyone who was anybody would simply call in at the EMS studio for a demonstration.

No expense was spared on making the studio comfortable to work in; it featured, for example, a carpeted and secluded "listening room." This was part of Peter Zinovieff's aesthetic—that electronic music ought to be listened to and created in an ambience free of machine noises. But the studio also shared Moog's friendly feel. The family atmosphere, with Peter's three children running around, made a deep impression on Robin when he first arrived there: "I remember going along, taking the 37 bus along the Upper Richmond Road and turning up in Putney and coming to this amazing place. And pop stars and people would regularly turn up . . . And all the composer friends that Peter and Tristram had and all their children . . . Yeah it was magic, really . . . It wasn't just a job . . . I was sort of part of the family."

At EMS lunches new projects were hashed out along with discussions of the latest studio techniques, manufacturing problems, gossip, and so on. Peter: "Whoever came to see us, whether it was an accountant, a lawyer, a manufacturer, somebody selling something, or the engineers or myself or other programmers, were just guests. We'd all have lunch together and sometimes it would be up to twenty or thirty people." These lunches provided a way not only of designing new products but also of showing off the company. It was Peter's view that if prospective clients sat down with them for lunch they were more likely to buy equipment because they had seen the "heart and soul" of the company. And indeed many of his top-of-the-line Synthi 100s were sold this way to visiting heads of electronic music studios in other countries, such as Radio Belgrade and Mossfilm, Moscow.

○

EMS and Musicians

Given Peter's lack of interest in pop music, Robin had to deal with the pop musicians who showed up at the Putney studio. This job suited him fine. Like nearly everyone else in the history of synthesizers, he had played in a group at school. The VCS3 and the Synthi AKS were very popular among progressive rock musicians, who mainly wanted to use them for their sound effects. Robin remembers that Dave Gilmour of Pink Floyd was a fairly regular visitor. Peter remembers attending one Pink Floyd concert with six VCS3 synthesizers on the stage. Pink Floyd was famous for breaking down the barrier between music and sound effects, and the group used EMS synthesizers on a number of its albums (for example, *Meddle*, 1971, and *Obscured by Clouds*, 1972). All four members of Pink Floyd are credited with using EMS equipment on the recording of one of the best-selling rock albums ever, *Dark Side of the Moon* (1973). Mark Cunningham in his history of record production comments: "An overwhelming use of synthesizers, especially the Peter Zinovieff-designed Electronic Music Systems [sic] (EMS) VCS-3 and Synthi A (a suitcased synthesizer with on-board keyboard and sequencer) provided a new range of sounds, none more sinister than on the instrumental 'On the Run,' for which the EMS sequencer provided the timing reference. Everything you hear on that track, apart from the sound effects, was done live by the Synthi A."[8] The fast-sequenced sounds of "On the Run," interspersed with sound effects of motor bikes and the like which build up in an explosive crescendo, produce the edgy, paranoid feeling of technology out of control.

Other well-known rock musicians who visited the studio included Pete Sinfeld and Robert Fripp of King Crimson. Many other rock and pop performers, such as the Rolling Stones and the Beatles, can be found on lists of EMS's customers, although what, if anything, they used the synthesizer for is unknown. Peter remembers trying to teach Ringo to play the VCS3, with little success.

One visitor to the EMS studios was Jon Lord of Deep Purple. There, he met "mad professor type" Peter Zinovieff: "I was ushered into his workshop and he was in there talking to a computer, trying to get it to answer back! He gave me one of their early models and I took it home to experiment with it, but I really couldn't get it to do much, it made odd bleeping noises, which wasn't terribly helpful to Deep Purple."[9] Lord went on to use the ARP Odyssey before discovering the Minimoog. Another leading rock performer to purchase a VCS3 was Peter Townshend, who used it to filter and sample and hold the organ on the well known introduction to "Won't Get Fooled Again," which was a huge hit single for The Who. Stevie Wonder was an EMS customer; Robin Wood flew to New York to deliver a sequencer to Malcolm Cecil and Bob Margouleff for use with TONTO on *Music of My Mind*.

One reason groups liked to use EMS synthesizers was because the audio

293

input allowed other instruments and signals to be processed by the synthesizer. Brian Eno's first synthesizer was a VCS3, and he used it to great effect by processing the other instruments in the art/rock band Roxy Music (for example, *Roxy Music*, 1971, and *For Your Pleasure*, 1973) and later Robert Fripp's guitar on the experimental album *No Pussyfooting* (1973). Eno, like most synthesizer players, uses several synthesizers, including the Minimoog, but the EMS does things the other ones cannot do:

> The thing that makes this a great machine is that whereas nearly all other synthesizers are set up so you have a fixed signal path . . . with the EMS you can go from the oscillator to the filter, and then use the filter output to control the same oscillator again . . . You get a kind of squiging effect. It feeds back on itself in interesting ways, because you can make some very complicated circles through the synthesizer. Also on the EMS every single function is on a potentiometer . . . Even the waveform is adjustable, as opposed to the Moog where you switch from one waveform to another.[10]

As with all great analog synths (and indeed all great instruments), the actual cause of the best sounds is shrouded in a degree of mystery. Robin Wood: "[the filter] has its own characteristic, and . . . although its got failings technically, even so it produces a characteristic sound, unlike anything else. Like if you overdrive it, it distorts in a certain way . . . The output channels overload in a certain way—they use germanium devices in the outputs and people say that germanium creates a very different sort of distortion than silicon."

○

No Musical Upbringing and Not Enough Money

There is also the matter of the VCS3's price. While price was almost certainly not a consideration for wealthy rock stars, there were plenty of other aspiring musicians who couldn't afford the more expensive Moogs and ARPs. For these musicians the VCS3 was perfect. Hawkwind, started by busker Dave Brock, became one of London's pioneering psychedelic space rock bands in the early seventies because of their use of synthesizers.[11] When two "electronic freaks," Del Dettmar and Dik Mik, joined Hawkwind, the VCS3 was all they could afford. A feature on Hawkwind in the London Underground newspaper *Frendz*, written in the disconnected prose and bizarre layout typical of the underground genre, nicely captures the times:

> It was Del the longest haired building labourer in the world, who entered with a hod on his back. And in the hod a bleeping, chirruping, wherping, blaspheming machine. "It's a sympathizer" he explained. "It must have heard the sounds coming from the room and started to sympathize. And now I can't stop it." Nik Turner [saxophonist] the birdman peered through the curtains. The room was curiously vibrating. "Hey" said Nik, "The room appears to be in outer space . . . unless my eyes deceive me . . . "The others looked. Dave Brock pulled back the curtains. "Right we are." He burst, "We are in space. We've taken off. What now???????" . . . Del and Dik Mik are both electronic freaks with no musical upbringing and not enough money to purchase a Moog.[12]

Armed with their "silver machines," Hawkwind took their audience on the ultimate intergalactic trip. The VCS3 was the most suitable of all the portable synthesizers for making psychedelic sounds. As Tim Blake, formerly of Gong and a notable solo synthesist, told Robin Wood, the VCS3 itself is "like acid."

◎

The Boys from Putney

The reach of the VCS3 soon extended beyond the shores of Merry Olde England. An American agent, Alfred Mayer, started marketing it in the United States as the Putney. He had his own company, Ionic Industries,

295

Figure 56. Hawkwind's Del Dettmar (left) at VCS3 and Dik Mik (right)

Inc., and even reprinted Tristam Cary's manual as an Ionic publication "By the Boys from Putney." Mayer managed to get some publicity for the Putney in U.S. newspapers, and it was launched at the fall 1970 AES, along with the ARP 2600 and the Minimoog. He aimed the $1,395 machine at the educational market and in one 1970 publicity sheet listed 27 different universities and colleges who had bought it.

The relationship with Mayer turned sour in 1972 when Ionic started marketing its own synthesizer, the Performer, with a manual written "By the Lads from Ionic." This synthesizer was suspiciously like a VCS3 (a "rip off," as Robin Wood put it) dumbed down to replace the patch board with a series of push-button controllers. The Performer promised "no requisites, no talent, no math, no physical development" to play it. The Performer also seems to have resulted in "no sales" and vanished without trace. Evertt

Hafner, a professor of music at Amherst College, replaced Mayer as the U.S. agent, and the Putney continued to sell steadily in the States, mainly to the educational market.

EMS also started to attend trade shows such as the Frankfurt show where EMS's agent in Germany, Ludwig Rehberg of Elektronik Musik Studio, arranged demonstrations of synthis for "a mixture of professors and a few sort of elite pop stars, you know people like Kraftwerk and Tangerine Dream, Klaus Schulze, those sorts of people." VCS3 synthesizers were taken up by many of these pioneering German synthesizer groups and were heard on a large number of their records from the early seventies.[13] In France, the well-known synthesist Jean-Michel Jarre bought a whole stack of VCS3s for his pop electronic music extravaganzas.[14]

The boys from Putney also started to attend NAMM shows, where they found the Americans treated them with a slight air of condescension. Perhaps there was an element of "Who are these upstarts on our territory?" Certainly with the high price of ARPs and Moogs in the UK, it was unlikely that the Americans would be able to penetrate the UK education market. In any case, by the seventies the American manufacturers were focused almost exclusively on performance keyboard synthesizers. Compared with keyboard synthesizers like the Minimoog, the VCS3 was found wanting; and of course, compared to the far more expensive modular machines, it looked extremely limited. When David Cockerell eventually left EMS and moved to the States to work on effects pedals, he found that few people in the music business had ever heard of the Putney.

○

The Fall of EMS

In understanding the fate of EMS, it is crucial to realize that the manufacture of synthesizers was not the original goal of the company. It was just something they did to support the main studio. And the main studio became increasingly expensive as Peter needed more and more advanced equipment. Early on Peter had fretted over what would happen to his studio in the long run. Private studios of this size were just not viable. The San Francisco Tape Center had faced the same difficulty before it got its Rockefeller Grant and moved to Mills College. In a letter he wrote to the London *Times* in 1969, Peter even offered to donate his £40,000 studio to the nation. Despite lobbying by Tristram and Peter, somehow it never happened.[15]

In the early 1970s EMS had employed 31 people, including 24 at the production facility in Wareham. In 1972 its sales turnover was 210,000 pounds and by 1973 18 Synthi 100s and 1,400 VCS3 and AKS model synthis had been sold. But in that year things started to turn sour. Two events coincided to make EMS's survival in its old form untenable: Peter met an entrepreneur on a flight to the United States, and his marriage with Victoria started to fall apart. EMS, built on Victoria's benevolence, had never had enough capital. The entrepreneur (who later hosted a famous American TV series) talked Peter into going into a joint venture that promised a huge injection of new capital. The idea was to turn the analyzer resynthesizer project into a new technology for telephony transmission. Peter's system with its analog digital conversion and reconversion showed promise as a way of compressing bandwidth and hence permitting more calls on the same cable. Peter was promised a Wall Street launch of the new venture.

In order to help EMS sustain the borrowings it needed to prepare for this public launch, the entrepreneur had arranged for EMS to be given a postdated check for 40,000 pounds guaranteed at an American bank (equivalent to over a million dollars today). With EMS's money shortage worsening by the day, its British bank finally decided in November 1973 to cash the American check, only to discover to everyone's consternation that it was a forgery. The whole deal fell through. This failure was a huge blow to Peter's morale and to the standing of the company.

At about the same time, Peter's marriage with Victoria also failed. She moved to a new house in Fulham, and EMS formally separated the sales office from the studio by opening a new sales office on a nearby street in Putney. The final blow came when Peter lost his resident genius, David Cockerell. David had been making his own sorties to the United States, and in New York he had met the owner of Electro-Harmonix, who offered him a huge salary to live there and design guitar effects pedals. It was an offer that Peter couldn't match and one that David couldn't refuse. At the end of 1973 David left for the States and later on to IRCAM in Paris to help Pierre Boulez build a new studio.

EMS soldiered on for a few more years, but the lunches were now not quite the same, as Peter was forced to worry about the wine bill. When the final marriage settlement came, Peter had to sell the house in Putney and moved the studio to a manor house outside of Oxford, the Priory. Sales of the VCS3 and Synthi A were starting to tail off as the Minimoog and ARP synthesizers were bought by more and more musicians. EMS did develop some new products. Its last gamble, the Polysynthi (designed by Graham Hinton), was a belated response to the Prophet-5. It never sold and had, by all accounts, a terrible sound.

In 1979 the company went into receivership. Today, Robin Wood, the man who joined EMS as a studio cleaner in 1970 and who has remained there through different changes of ownership, runs what is left of the company from his remote cottage in Cornwall.[16] He still makes and sells about forty EMS synthesizers a year (VCS3 and Synthi A), mainly to pop groups. The design is virtually unchanged from thirty years ago. In 2000 Oasis bought one and Radiohead bought two.

◎

EMS *versus Moog and ARP*

EMS, unlike Moog and ARP, was a cutting-edge research studio that happened to develop a design and manufacturing arm to help support it. Cockerell, its engineer in chief, was located elsewhere. He would visit the studio about two or three times a week and Peter would conduct long conversations with him by telephone from his bath tub, but by and large David did not meet the musicians who turned up at the Putney studio. This helps explain why EMS did not develop what most musicians wanted in the 1970s—a keyboard performance synth. When he arrived in New York to make effects pedals for Electro-Harmonix, David found a much closer interaction between engineers and musicians: "I learned then that one's got to listen much more closely to musicians. We had many musicians at the Electro-Harmonix factory, and in fact the guys who tested the machines, the guys who sold them, they were all, many of them were really wonderful musicians."

Here we can point to the wider culture and its role in shaping the synthesizer. Peter Zinovieff, the son of an aristocrat, ran his studio like an upper-class salon—in effect there was no place for "trade" there. Trade—engineering and manufacturing—were carried out elsewhere. But trade is where the money was (once his wife's supply of jewelry was exhausted) and where David and the engineers were. The schisms of Britain's famously class-ridden society were played out in the doing and undoing of EMS.

©

End of a Dream

Water has always been a part of Peter Zinovieff's life. He loved his Scottish island retreat where he went fishing and sailing, and his favorite studio at EMS was the one overlooking the mighty Thames. But it was water that finally took his beloved studio. Through his personal connections with Harrison Birtwistle (now Sir Harrison Birtwistle) and the good offices of Sir Peter Hall, the National Theatre agreed to store Peter's studio after the bankruptcy. Peter: "They stored it in the National Theatre in a dungeon, and it rained on it . . . that was the end of it." Robin Wood visited that "dungeon": "I remember going back there with Peter and seeing it all in bits. It was an awful sight, all this stuff, all just sort of lumped together in this huge room." As the water seeped in, the equipment slowly rotted; eventually it was all junked, apart from the Ampex tape recorders. Peter's dream was over.

The legacy of Zinovieff, Cockerell, and Tristram Cary lives on with the extraordinary VCS3 synthesizer. It may have been a byproduct of their other projects, it may have been seen by many as a toy, but it introduced lots of people to the ideas of electronic music and it made some of the most inventive and memorable rock music of the 1970s. The little synthesizer had a lasting impact. Its siren sounds also led to the writing of this book.

Conclusion: Performance

ONDON AIRPORT, the summer of 1968. The Man from Moog (Jon Weiss), laden down with black musical equipment carrying cases, stepped off the flight from New York. Allen Klein, then manager of the Rolling Stones, famous for striking a tough bargain, had made a special deal with Moog whereby the Stones would receive not only a Moog Series III modular synthesizer but also a week's free tuition. Jon was also delivering a pair of JBL speakers which the Stones had requested he bring from the States. These speakers, unavailable in Britain, were perfect for playing loud, distorted rock 'n' roll.

When Her Majesties Custom and Excise Officers learned that the mysterious black boxes were for delivery to a certain Mr. Jagger, they spent three hours taking them apart searching for drugs. Jon, who is the mildest guy in the world, told us with typical understatement, "They were not very nice." Welcome to Swinging London!

Jon, a classically trained violinist, owned just one Stones album and didn't know for sure which Stone Mick Jagger was until the London trip came up. His lack of expectations almost certainly helped him befriend Mick, who, Jon was pleasantly surprised to discover as time went on, was very different from his image: "It was just outrageous. The public image . . . and then once you're inside, at home, this guy had the most varied record collection I'd ever seen . . . Of course, he had all the old American blues albums, but he had classical stuff, he had avant-garde, serious music—I mean, he listened to everything."

On hearing the acetate of the Stones' latest LP, *Beggar's Banquet* (1968), Jon became puzzled. The R&B format of the album marked a return by the Stones to their roots after the psychedelic experimentation of *Their Satanic Majesty Requests* (1967). He had assumed that the Stones had bought the synthesizer to make more psychedelic music, but now they had gone in a very different musical direction.

Exactly what they were going to use their newly acquired synthesizer for was not yet clear even to the Stones. Jon told us Mick (with hair dyed almost black) was playing with the idea of using the Moog as *his* instrument in the band. He was also in the middle of making a movie, *Performance*, and the synthesizer could be useful for the score on which he and Keith Richards were working. Jon patched Keith's guitar through the Moog. Although Keith liked the sound, the control just wasn't there (pitch-to-voltage converters came later). Bill Wyman and Brian Jones complained that Mick was hogging the new toy to himself.[1] Jon went through his paces, explaining the instrument; he found that Mick was a proficient learner: "Of all the people that I ever showed how to use the machine, he picked it up the fastest. He's just an *incredibly* sharp guy."

As one week turned into another and then a month, Jon (living in Mick's house) became close to Mick and part of the Stones' inner circle. They hit on the idea of using the synthesizer as a prop in *Performance*. Mick was playing the part of Turner, a fading rock star, who lives in a freak house in London surrounded by weird and beautiful people. Turner has a mysterious recording studio in the basement bedecked with Indian rugs and sixties gewgaws. The Moog with its rows of knobs and dials would make a perfect addition. Jon took the synthesizer onto the set, much to the amusement of the English workers, who had never seen the "fabulous sanitizer" before, and set up a little patch for Mick to dabble with.

Performance, written by a painter friend of Jagger's, Donald Cammell, and co-directed by Nicholas Roeg, is a strange but brilliant movie. Warner Brothers sunk 1.8 million pounds into it, expecting something like the Beatles' *Hard Day's Night* (1964). Filmed in 1968, it was not released until 1970 and then only in a heavily edited version. Another *Hard Day's Night* it most certainly was not. It was laden with counterculture values—sex, drugs, rock 'n' roll, as well as cabalism and a bizarre magic and drug-induced transmigration between the two main characters, Turner and a London gangster, Chas (played by James Fox). It became a classic sixties cult movie, and its clothing is still a source of inspiration for New York fashion designers.

Chas, on the run from his gang, inadvertently stumbles upon Turner's freak house and is offered shelter there by Turner's "secretary," Pherber. In real life, the actress playing this part was Anita Pallenberg, the beautiful German model who Brian Jones had first befriended and who, at the time, was Keith Richard's girlfriend (Mick's sex scenes with Anita were a source of some tension among the Stones). Chas and Turner become infatuated with each other. Turner's muse has deserted him, and he wants to perform again. Chas is a performer—the gangland term for a hired killer. Chas, increasingly disillusioned with his gangland friends, slowly falls under the influence of Turner and *his* gang (Pherber and the young girl, Lucy, who form Turner's *ménage à trois*).

In the key scene, Chas is summoned to meet Turner. Dressed in outlandish clothes, body painted and tripping on magic mushrooms, he enters the mysterious studio to find Turner squatted on the floor playing his Moog synthesizer. A tape recorder plays menacing music ("Poor White Hound Dog," a blues with slide guitar and vocal by Merry Clayton). As Turner puts the finishing touches to his patch, two figures (Turner and Chas) can be seen dancing in the background. All is a wall of mirrors. Patch complete, Turner grabs a fluorescent light bulb (with cord just visible—the special effects left something to be desired) and starts dancing with Chas. Turner's increasingly frenzied dancing peaks with a final thrust of the light bulb toward Chas's ear.

A moment of sixties freakout follows: psychedelic swirling colors, images of spinning keyboards, close-ups of Chas's ear (with all the sexual overtones imaginable), and a sinister drone from the Moog. We appear to be looking through a tube (the light bulb, a reverse telescope, or a worm hole?) and the scene switches to a London club and there is Turner with Brylcreamed, swept-back hair but now singing an R&B song ("Memo from Turner") to all of Chas's underworld cronies. Turner (or is it Chas) is at last performing again.

This is the only movie we know of where the Moog synthesizer itself makes a cameo appearance.[2] Turner for a moment is the mad captain at the controls of spaceship Moog. The Moog and its sounds are the perfect prop, part of the psychedelic paraphernalia, the magical means to transmigrate a fading rock star into something else.[3]

The Moog was a machine that empowered such transformations. The synthesizer for a short while in the sixties was not just another musical instrument; it was part of the sixties apparatus for transgression, transcendence, and transformation. No wonder the sixties rock stars loved their Moogs.

In the end, Jon returned to Trumansburg, where two years later Bob Moog was forced to sell his business; Mick did not take up the synthesizer, but that particular Moog synthesizer lived on.[4] It was sold on to the Hansa by the Wall recording studio in Berlin, where in 1973 Christoph Franke of Tangerine Dream purchased it for $15,000. The Moog sequencer became the defining element of Tangerine Dream's sound, and the Moog became an enduring influence on the many waves of German electronic music during the 1970s.[5] This influence eventually provided renewed stimulus in the United States when Donna Summer's *I Feel Love* (1977), produced by Giogio Moroder in a Munich studio with the aid of a modular Moog, along with Kraftwerk's *Trans-Europe Express* (1977), were taken up in black

305

dance culture (particularly Detroit techno) and led to the explosion of synthesizer-based dance music in the eighties and nineties (more of which below).

○

In Court

It was 1994, and Bob Moog and Herb Deutsch were together again. They had both been called as "expert witnesses" in a lawsuit between the giant Japanese company Casio and the U.S. Government.[6] Casio wanted to pay a lower tariff on the so-called synthesizers it imported into the States—the thousands upon thousands of little keyboard devices that produce simple musical sounds in fixed preprogrammed patterns and lurk in kids' bedrooms everywhere.

What the court had to decide was whether the Casio devices were machines or musical instruments. Casio wanted them classified as "electrical articles" and hence subject to a lower tariff than that leveled on musical instruments. It fell to Bob, testifying for Casio, the plaintiff, to argue that the Casios were machines and to Herb, testifying for the government, to argue that they were musical instruments. Bob pointed out that the essence of a musical instrument is that the performer should have "real-time control" and that the Casio takes this control away. The judge did not buy this argument, describing it as a "seemingly myopic premise" and contrary to legislative intent. The government won the case and the judge decided that, for tariff purposes anyway, the Casio, as long as it contained an amplifier and loudspeakers, was a musical instrument and thus subject to the higher rate of tariff.[7]

The court was grappling with an issue that not only marks a fault line in the world of synthesizers but also has been debated throughout the history of music. Whenever a new mechanical contrivance enters the field of music, it triggers the same set of concerns: is it a proper part of the musical domain, an instrument that can release musical talent, creativity, and art, or is

306

it simply a mechanical device, a mere machine? This debate famously occurred when the piano forte replaced the harpsichord at the start of the eighteenth century. Much less well known is the similar debate that arose when mechanical levers were added to flutes in the mid-nineteenth century.[8] The linked key mechanisms and valves that replaced the use of fingers over individual holes were found to be easy to operate and facilitated the production of much more uniform and cleaner tones. But opposition came from those who found that the new keys ruled out the possibility of making a "vibrato by simply moving the fingers over the sound holes" and meant a loss of control over finger positions "to correct out of tune sounds."

It seems that there are always people who oppose the incursion of new mechanical devices into music. The boundary between musical instrument and machine is continually being redrawn with each new encroachment. At the turn of the twentieth century, the piano trade labeled the idea "that such a thing as a 'machine' [could produce] piano playing as 'ridiculous' and 'preposterous.'"[9] Musicians, music teachers, and composers opposed the player piano, stressing that one could copy sound but not interpretation; that mechanical instruments reduced the expression of music to a mathematical system; that amateur players would disappear; and that mechanized music diminished the ideal of beauty by "producing the same after same, with no variation, no soul, no joy, no passion."[10]

Other musicians and composers welcomed the new mechanical instruments, claiming that they had the potential to replace expensive musicians, were better and more precise as performers of the ever increasing complex music, and had more possibilities to express the objectivity of the unfailing precision and collective spirit of the age.[11] Others similarly stressed the need for clear and unsentimental music or saw the specific potential of mechanical music when composed for radio and gramophone—technologies more suited to rigorous, linear, and rhythmic music. Another argument came from supporters among American music educators, piano manufac-

CONCLUSION: PERFORMANCE

turers, and music publishers, who thought that the player piano (which was for a while a hugely successful instrument) would lead to "an almost universal music education" and therefore "democratize music."[12]

The debate surrounding the synthesizer is simply the latest twist in these old arguments. A product of the collaboration between engineers, musicians, and salespeople, the synthesizer throughout its history has been adopted by different communities for very different purposes. For some it is an archetypical machine, a way to abstract and analyze the core constituents of sound, a way to render into algorithmical form, and manipulate and perfectly repeat with machine-like precision, the essence of sounds and music. For others it is a musical instrument with all the idiosyncracies and inaccuracies associated with the best acoustic instruments—an object to love, to learn to work with, to appreciate for its ineffable qualities and its own personality.

◯

Liminal Entities

The careful reader will notice that we have avoided taking a position over whether the synthesizer (or flute or player piano) is solely a machine or solely a musical instrument. Rather than read any essence into this technology, we have looked at what people themselves make of it—the meanings they derive from its use. The synthesizer is a form of "boundary object," a liminal entity. Liminal entities are "neither here nor there; they are betwixt and between the positions assigned and arrayed by law, custom, convention and ceremony," according to the anthropologist Victor Turner.[13] The synthesizer is something that can pass between different worlds, that can take on different meanings in these worlds and in the process transform these worlds.[14] The question of whether the synthesizer is a machine or a musical instrument, of whether it is for classical or pop music, of whether it is for emulating old sounds or exploring new ones, of whether it is a part of science or art, will never be adequately answered. They are the wrong questions to ask. There are many souls in this new machine.[15]

Our approach draws attention to the different cultural meanings and social worlds woven around the synthesizer by its *users*. Technologies are never neutral; they are always embedded in and generated by a cultural context, and the most important cultural context is that of use. When synthesizers first appeared, they were equally at home in the electronic crucible of the Trips Festival, in a studio making Bach, or on stage before 10,000 young fans at an Emerson, Lake and Palmer concert.

Understanding how meanings get woven into technologies by users is the key to solving one of the central puzzles of our book: why Moog's conception of the synthesizer prevailed over Buchla's. The crucial moment occurred in 1964 when Moog first hit upon the volt-per-octave standard and built his exponential converter circuits and then the rest of his synthesizer around it. It looked like a fine, neutral, sensible, technical standard that was soon adopted by ARP and EMS, but such standards, like machines themselves, are never neutral.[16] What he in effect did was to embed into his technology a piece of existing culture—the idea that music is about intervals. By defining octaves, the Moog preordained the keyboard as the controller of the synth.

But it didn't have to be this way. Don Buchla had a synthesizer design that didn't follow the volt-per-octave standard, didn't have keyboards, and didn't invite conventional melodic music. But by *not* catering to such music, Buchla had boxed himself him, because playing that sort of music is what most users wanted to do. Within five years the game was over. Moog's innovation had become the Minimoog—a keyboard instrument—and Buchla was a neglected hero making his instruments for the vanguard and exploring his own art.

○

Technological Frames

"Technological frame"—like "paradigm"—is a term that captures the way a whole series of practices, ideas, and values get built around a technology.[17] It includes both the ways technologies are produced and the ways

they are used and consumed. Moog's technological frame was to mass produce and market a well-engineered, reliably serviced product that was re-
sponsive to the needs of users. Buchla, on the other hand, had his own very
singular vision, and although he too learned from the musicians he met,
he never did this to the same extent as Moog. The sorts of musicians that
used his instruments were often like-minded members of the avant-garde.
Buchla's own musical sensibilities were both his strength and his weakness.
He knew what electronic music composers wanted because he was one
himself, but by refusing to accommodate to the needs of other users, the
Buchla technological frame was largely limited to a fringe market. Being
an experimentalist and artist by temperament, Buchla was quite happy not
to serve the commercial world.

    Moog, by temperament, was not completely comfortable with the busi-
ness world either. This goes back to his early experiences at Bronx Science,
where he felt alienated from the other kids who had wealthy, business-ori-
ented parents. Moog's company was thus never a fully viable commercial
enterprise, as ARP would later become. There always was a quirky feel
about Bob and his funky factory. This home-spun touch actually helped fa-
cilitate the interaction between musicians and engineers, but in the long
run Bob's lack of business acumen was fatal for his enterprise.

    The different technological frames of the two synthesizer pioneers were
shaped by the wider culture within which they both worked, and both in
turn shaped that culture. Buchla's frame emerged within the artistic milieu
of the San Francisco Tape Center and was shaped by the composers he met
there and the sixties counterculture of which he and the Tape Center were
a part. Moog's frame, on the other hand, was shaped by conservative fifties
engineering values and the lower-middle-class Protestant work ethic that
predominated in his Trumansburg factory. Moog was not an elitist. Indeed,
his focus on new users, like the marketing of the player piano, meshed well
with the idea that music should be available to all. Later Moog was to express his enthusiasm for the democratic possibilities of the synthesizer as,

in its cheap digital form, it became available to many more users.[18] Although his own synthesizers were too expensive for most users, he set the synthesizer on a path that led to a much larger user base.

Moog was from the same immediate postwar generation as Buchla, but he never became a sixties person. The sixties came very late to Trumansburg, and he was more a bemused bystander than a participant. This is a case where geography really mattered. The acidheads in San Francisco seemed a million miles away from life in upstate New York.

But none of this was predetermined. It is easy, with hindsight, after the keyboard synthesizer became the dominant form, to talk about Buchla's synthesizers as being peculiar or different. At the time, Buchla's vision for the instrument was in every way as legitimate as Moog's. What we have tried to do is show how this technology, which in 1964 had two possibilities, two different meanings, and considerable "interpretative flexibility," eventually took the predominant form that it did.[19]

○

The Culture of Use

Our story of the synthesizer draws attention to the role played by users.[20] Designers "script" or "configure" ideal users into their machines.[21] The black and white chromatic keyboard scripted a certain sort of user: one who wanted to play conventional melodic music. Scripts try to constrain the agency of users, but users can exert agency, too, and can come up with their own alternative scripts. Hip-hop DJs use turntables for "scratching," a use inconceivable to the engineers who first designed them. It was possible to tune Buchla's touch pads to the conventional scale and to retune Moog's keyboards into unconventional scales, but such reconfigurations — the making of new scripts — required specialized skills, and most users did not want to invest the time and effort.

Users do not come to technology unprepared. They are part of a wider culture of use, and they learn within that culture. They may have invested years of practice on conventional keyboard instruments and their listening practices may have been subtly shaped by the dominant genre of melodic music. As in the case of the QWERTY keyboard on the computer (named after the first five letter keys on the left top row)—a cumbersome design first used on the typewriter to slow down typists and then taken over for computers—it is no easy matter to learn a new keyboard layout, especially when all the machines in the environment and the culture of use are predicated on the QWERTY system.

Analog Days shows that the worlds of synthesizer production and consumption were not separate worlds.[22] What our story reveals is that consumers have played a crucial role *throughout* the history of the synthesizer and in all aspects of its development, including design, testing, sales, and marketing.

The *design* of the Moog modular synthesizer came about from collaboration between users and engineers. Musicians like Herb Deutsch, Walter Sear, Wendy Carlos, Eric Siday, David Borden, and Paul Beaver helped shape the design of the synthesizer. And then they acted as a *test* laboratory for Moog's new prototypes and products. It was out of the interaction with users that the archetypal synthesizer, the Minimoog, was born. The engineers were surrounded by musicians at the factory; musicians took the early Minimoog prototypes out to road test and were a constant source of feedback.

But the impact of users did not end once Minimoogs began to leave the factory. The *selling* of the Minimoog is also a story about the role of users. David Van Koevering's use of first the modular Moog and later the Minimoog enabled him to grasp its potential for a new audience of rock 'n' rollers. He developed not only a new way to sell the instrument but also methods to make sure that musicians could use it—he saw the importance of sound charts and an instruction manual. When he became vice-president of Moog, he was able to implement many of his ideas about how this instrument should be sold on a global scale. The study of salespeople has been neglected.[23] Yet they are a crucial link between the worlds of production and consumption. Whether through their interactions with users or by moving from use to sales, salespeople tie the world of use to the world of design and manufacture. It is their mobility—moving between different users and among different networks or back and forth between manufacturers and users—that makes them key mediators to study in the development of technology. The salespeople bring into being not only a new market but also new alignments between manufacturers and that market, helping to change the culture of use.

In *Analog Days* we have seen how a new *market* for synthesizers was born. Economists tend to talk about markets purely in economic terms, as the places where supply and demand meet.[24] The new market for synthesizers did not sit out there waiting for the right product to come along—it had to be actively created. This is the case with so many successful products. For example, George Eastman designed the first portable film roll camera—the Kodak—in 1888; but he had to turn photography from an activity carried out by small numbers of professionals to one where everyone could participate. In the process he actively recruited people to his new vision of popular photography. In the case of the synthesizer, as with the Kodak camera, designers, sellers, and users all played key roles in bringing a *new product*—the Minimoog—sold in a *new way*, to a *new group* of users.

◯

Boundary Shifters

Analog Days shows the mirror dance between humans and machines in the course of technological innovation. Just as synthesizers transgressed and refused to keep their identities within bounds, so too did the humans who used them. The identities we have assigned to the actors in *Analog Days* are in an important sense inadequate: categories like engineer, musician, and salesman are constantly being called into question as actors refuse to comply with the labels that we analysts give them.[25] This blurring of

313

categories seems an integral part of the transformation we have been studying. When the modular Moog synthesizer was first used in recording studios, no-one knew what to call its operators: were they engineers, programmers, producers, musicians, or what? The longer term shift in categories brought about by synthesizers, samplers, and concomitant changes in studio technology was even more profound: what counts as an instrument, what counts as a studio, what counts as a composition, and what counts as live performance all eventually undergo transformation in a revolution whose effects are still being felt.

We need new ways to designate not only the liminality of machines but also the liminality of the human roles and identities built around the machines.[26] We need to describe, for instance, how actors can slip the anchors that keep them tied to just one identity and how new identities themselves come to be. Not only do people change identities, transgress boundaries, and move from one world to the other—say, from engineering to music— but they also apply the knowledge, skill, and experience gained in one world to transform the other. Thus, a Bob Moog morphed back and forth between his engineering world and the world of musicians and in the process he transformed the synthesizer. We call such people "boundary shifters"—people who cross boundaries and in so doing produce a transformation.[27] For an organization successfully to innovate, it must allow for such boundary shifting. Salespeople would seem to be quintessential boundary shifters.[28]

○

Psychedelia

To mass-market synthesizers, Moog needed to find new users and the synthesizer needed to escape from the world of the avant-garde. Paradoxically, this new use emerged first within the electronic melting pot of the Trips Festival and who should have discovered it but Don Buchla. The electronic sounds of the synthesizer blended perfectly with the other mind-altering explorations going on at the same time. The technology of drugs, light shows, and sounds merged together into the world of the psychedelic sixties.

On the West Coast at the Monterey festival, the search for new psychedelic sounds and a synthesizer that could be commercially produced came together. Critical to this meeting was the record industry, whose contracts enabled new rock artists to buy expensive synthesizers. At Monterey, Moog reaped the harvest of what Buchla had sowed, and the new psychedelic sounds of the synthesizer started to become part of, and to transform, the culture. Moog's synthesizer became *the* synthesizer. This early success came from the desire of two customers, Paul Beaver and Bernie Krause, to act as his sales reps. This odd couple saw the potential to recruit new users. They too were boundary shifters. They saw how the Moog synthesizer was the perfect accompaniment to cash-rich, spaced-out rock musicians. In the process they sold synthesizers, played sessions, trained new users, formed networks with customers, offered Moog advice about how to improve his product, and produced some of the most memorable music of the era. Having been transformed by this new group of users, the synthesizer and its sounds further transformed listeners, allowing them to enter new psychedelic soundscapes.

The liminal status of the synthesizer is most apparent when its operators themselves crossed boundaries and transgressed social worlds. Women synthesists broke the mould in more ways than one. Suzanne Ciani found a new analog identity that was as ambiguous in its own way as Walter/Wendy Carlos's—a new sort of machine-person hybrid.[29] It was not only the identities of machines and humans that were in flux but the very boundaries we erect between the two categories. When these woman synthesists took on an identity associated with a machine, and in Suzanne's case feminized the identity of that machine, we see that another boundary has been transgressed. Here, at this point, it is our notion of the machine that needs to be changed—it was not for Suzanne a cold lifeless thing to which she was a slave but a breathing, living partner in an evolving relationship. By confessing her love for a machine, Suzanne reminds us that no boundaries are sacred and shows us the potential for new sorts of human-machine couplings built around this technology.

By the time Carlos released *Switched-On Bach*, the ground had been laid. The culture was ready to cross over. Maybe not ready yet for Carlos's own personal crossover, but ready for Bach to be psychedelicized, for Bach to be switched-on. The beauty was that it was Bach and not "What's New Pussycat." Carlos's album allowed a whole new audience to experience the rush of the sixties without having to smoke dope, engage in radical politics, or listen to loud rock music. All they had to do was put a bit of Bach on the turntable. In the process the ultimate users, the listeners, allowed their boundaries to shift—they had effectively been recruited to a new social world by a piece of technology (mediated of course by a best-selling record). Electronic music was now a normal part of the soundscape, and listeners had been changed forever.

We usually think of the counterculture as anti-technology but the new technologies of sound and light, combined with mind-altering drugs, were an integral part of the movement.[30] The counterculture was a breeding ground for many new cultural artifacts and processes, from tie-dyed shirts to geodesic domes.[31] One message of *Analog Days* is that the technologies and sounds of the counterculture were an (unlikely) source of today's mainstream digital audio culture.

◎

Digital Days

Since the pioneering work of Moog and Buchla, the synthesizer has evolved almost beyond recognition. We are now in a digital world where sounds are produced in bits on digital computers and processors. Digital synthesizers and a new instrument, the digital sampler, are commonplace in today's music.

The transition to digital has been complex and involves many different people and many different instruments. It includes the establishment in 1981 of a new standard, MIDI (Musical Instrument Digital Interface). The Yamaha DX7 produced in 1983 is usually regarded as the breakthrough digital instrument, the first one to achieve commercial success. The DX7 sold 200,000 units in three years, compared with the Minimoog's 12,000 lifetime sales.[32] And with the DX7 an important change in the whole field of synthesis took place. Although the DX7 was programmable, Yamaha found that synthesizers returned for repair contained almost exclusively the factory sounds they had been sold with.[33] The complexities of programming, compared with the ease of use of the factory pre-set sounds, meant that users of the synth either no longer wanted to or were unable to explore and find new sounds. Soon a secondary soundcard industry evolved that programmed special sounds for the DX7.

Four characteristics were shared by most digital synths. First, the built-in keyboard became ubiquitous after the success of the Minimoog. Second, the sounds were invariably accessed and controlled by means of a digital menu of pre-set buttons.[34] Third, the sounds themselves were pre-set and included emulations of acoustic instruments, emulations of other electronic instruments (including the Hammond and Moog), and completely new sounds with made-up names. Fourth and last, the technology was self-contained and much harder to modify and customize. We have seen how all these characteristics started to develop in the analog phase.

◎

The Analog Revival

Although we live in the digital age, there is something enduring (not to say endearing) about analog synthesizers. Today, an analog revival—a return to "knobs and wires"—is in full swing.[35] It has taken several forms. Old instruments, which now command record prices, are much sought after by modern musicians. At the biannual NAMM show, vintage synthesizers are demonstrated and are increasingly being taken up for live use and in the recording studio.[36] And digital synthesizer manufacturers, ever sensitive to the market, are offering more knobs on their newest models.[37]

It is easy to dismiss this analog revival as a form of nostalgia. Nostalgia is usually taken to be a means whereby present uncertainties and discontents are addressed by drawing on a past era or culture.[38] We get nostalgic only when we are having a problem with the present. Certainly it is easy to romanticize the sixties and to treat an interest in sixties technology as part of a yearning for the values of the peace and love generation and the definitive music it produced. But we think something more interesting is going on. In users' adaptation of and reversion to old technologies we see salient criticisms of how the synthesizer has evolved and expressions of genuine feelings of loss.[39]

The synthesists whose stories we have told in this book feel this loss acutely. For many, the reason they got excited about the synthesizer in the first place was because of its vast range of sounds. Some, like Bernie Krause, bemoan how the synthesizer over time has become more and more limited. What Krause observed in the LA studios was the effect of the wider culture (specifically the culture mediated by the record industry) on the synthesizer. Only certain sounds could be recognized, described, and communicated, and those sounds became embedded in the technology, first with sound charts and later pre-sets, thereby reinforcing the recognizability and reproducibility of these same sounds.

Many synthesists and engineers express the sentiment that somehow the synthesizer did not evolve as they wanted or expected it to. They sense a missed opportunity, a technology that slipped through their fingers without being exploited to the full; as if, in the rush to digital, something important about sound was overlooked. As Brian Eno has commented, "If I built a synthesizer, it would be fairly unpredictable . . . that's one of the things I feel is missing with synthesizers—a personality."[40]

Eno is expressing a widespread desire among synthesists for a real musical instrument, something imperfect, a living-breathing entity that you can interact with and even fall in love with. Most important of all, there is a desire for something that can discover the interesting sounds—sounds "between the knobs," as it were. For some people, digital sound is too perfect, too clean, too cold—they long instead for the imperfections of the warm, fuzzy, dirty analog sound.[41]

For many of the people we interviewed, the modular Moog synthesizer and the Buchla Music Box were just this kind of real instrument. Jon Weiss: "There were . . . certain inaccuracies in the equipment that resulted in wonderful and bizarre events . . . in that sense it was an instrument, it wasn't a machine. A machine would have created no inaccuracies and I think that's maybe why these computer digital generated sounds are not as interesting as the analog sounds . . . Accuracy like that doesn't exist in our lives, nature is never accurate, there are always weird concussions of sound waves, and overlapping and so on."

◎

The Synthesizer Nuts

David Cockerell, the man who built the VCS3 and who today builds digital samplers for Akai, is amazed that people still buy old synthesizers. "I love the way there are synthesizer nuts, the way there are classic car nuts . . . I don't understand it. I'm fascinated with sounds. Crummy old machines that aren't half as good as a ten dollar Casio, I don't understand why anyone would want them."

Brian Kehew, a young musician who plays with Roger Manning in the LA band Moog Cookbook is someone who does want old synthesizers. Brian *is* a synthesizer nut and has a huge collection of analog and digital synthesizers. He plays them in his own band, on studio sessions, and sometimes live with other groups, such as the French synthesizer group Air. He lends synths to fellow LA musicians like Beck (with whom Roger Manning also plays keyboards).

Amazingly, Brian got interested in synthesizers through *Switched-On Bach*, which he first heard in grade school on his mother's car radio. Brian started collecting analog synths in the 1980s when they were still cheap: "I really thought the control and the sounds of the analog synthesizer were much better than the new keyboards that had one slider, or a data entry. I thought that was a horrible way to do anything. I love digital synthesizers to this day—I like anything that creates sounds—but you have to be able to control it, otherwise there's no fun in it for me." For Brian, "The more knobs, the more fun." It is the vast range of sounds to be found in the analog instruments that interests him. The lack of exact reproducibility doesn't really bother him because "there are more sounds that I've never heard yet."

The pop keyboard synth music of Moog Cookbook makes knowing references to the switched-on genre of records, and to Perrey and Kingsley and their cheesy synth solos. Air's album, *Moon Safari* (1998), is a hypnotic collection of references to earlier analog synth sounds and even features the legendary Jean-Jacques Perry on Moog. The analog revival features numerous other bands and performances like Stereolab, Nine Inch Nails, Radiohead, Moby, and Tortoise, but there is also enormous interest in the analog sounds in many different genres of music. In rap, hip-hop, and the multiplying genres of electronic dance music, young musicians armed with digital samplers can take sounds from the analog world and reuse them in new contexts.[42] Analog still lives on in a digital world.

The interesting question remains as to whether the synthesizer and electronic sound in the digital age will continue to lead to moments of transgression, transcendence, and transformation. The answer is yes. The same chemistry of sound discovered by Ken Kesey, Don Buchla, Ramon Sender, and others lies at the core of electronic dance music and especially the raves and dance parties that have been sweeping the planet since 1988.

Sizzle and Boom

Rave was a term first used in the sixties to describe communal "happenings" such as the rave at the Roundhouse in 1967 where Peter Zinovieff's group, Unit Delta Plus, and Paul McCartney supplied the electronic sounds. Today's raves are communal dance experiences, attended sometimes by thousand of ravers. With a powerful sound system, a light show, and a skillful DJ who responds to and can work the mood of the crowd, the ravers dance all night to some brand of "house" music to achieve an Ecstacy-driven state of communal bliss. Ecstacy (the psychedelic amphetamine 3,4 methylenedioxymethamphetamine, or MDMA), often simply known as "E," is the new drug of choice. Often a rave includes chill-out areas where the more mellow genres of "ambient house" and "New Age house" music are played to help maintain a psychedelic quality to the coming-down phase of the Ecstasy high.

The genesis of electronic dance music is complicated, as is its multifarious forms, and both lie beyond the scope of this book.[43] But there are several obvious parallels between today's e-dance music and the sixties psychedelic music. The illicit nature of the early acid house parties and raves have led to comparisons with the earlier sixties counterculture. The crescendo of great outdoor raves held in the English countryside in the summer of 1988 is often described as the "second summer of love." The participants at the early rave events were not averse to making direct links back to the sixties. Witness this description of a "Spectrum: Theatre of Madness" event staged in spring 1988 in a London Club called Heaven and advertised with posters that drew directly from designs used by the Grateful Dead in their heyday: "The wide green beam of Heaven's lasers captured the outstretched arms and convulsing fingers reaching through the dry ice . . . 'Can you pass the acid test?' demanded the flyers."[44]

Some of the instigators of raves, like the Spiral Tribe, adopted the same

321

nomadic lifestyles as the early hippies: "Like Ken Kesey's Merry Pranksters . . . the Tribe grew as it left a kaleidoscopic trail through the idyllic countryside."[45] Sheila Whiteley notes: "There is a strong sense of shared identity between the sixties hippy philosophy and that of nineties alternative culture. Similarities are present in the music, the influence of drug experience (LSD/Ecstacy), an awareness of destruction and ruination of the earth and the poisoning of the seas . . . Collective experience, music and drugs appear, once again, to provide the means whereby young people can explore the politics of consciousness, to set up an alternative life style."[46]

Of course, the similarities can be overstated. Eighties hedonism has replaced much of the political sensibility of the sixties generation. The "right to party" has been perhaps the e-generation's best known political credo. Where the similarity is strongest, however, is in the role of electronic sound and drugs acting together in a communal context to produce transcendent experience. Again, as Sheila Whiteley has noted, "The atmospheric textures and multi-layered spatial compositions of the sixties psychedelic music produce a similar effect to the techniques for the manipulation of recorded music used in both House and Ambient music where, editing the start/end of the sample, repeating (looping), reverse and Low Frequency Oscillation and velocity sensitivity are all integral parts of the mix."[47] For example, the ambient trance music of the Orb (*Adventures beyond the Ultraworld*, 1991) uses a modular Moog synth and evokes a spacey disconnected sound reminiscent of early Pink Floyd.

Many of today's e-music groups, DJs, and remixers use analog synths, often along with digital samplers. Derrick May and Frankie Knuckles' experiments with a Roland analog TR-909 drum machine is said to have led to techno and house music.[48] Knuckles is reputed to have bought the Roland from May and used it to "segue between tracks and to crank up the sound of the bass kick at a crucial point in the song." A whole generation of Roland analog bass and drum machines like the Roland TB-303 bass machine are core constituents of techno, and DJs and techno groups even name themselves after these early Roland machines. As one commentator notes: "Using drum computers marketed by Roland of Japan in the early eighties which by this time were obsolete, discontinued and available cheaply on the secondhand market, Chicago's young hustlers wrenched out the possibilities that the manufacturers had never envisioned . . . their sizzle and boom locking into the mood of the clubs. Over a big sound system they reverberated through flesh and bone. This was do-it-yourself music; anyone could join in . . . you could just fire up your box and go."[49] This is a sentiment that Dennis Houlihan, the president of Roland USA, would applaud. He once told us he wakes up each morning and says, "Thank God for rave!"

The story of users thus continues for this new generation of sonic-hackers. By adapting and changing old analog technology and using it as it was never intended to be used, and by combining it with new digital techniques, they have found a way once more to shape the technology and to revive the alchemy of sound. The analog days are here again with a vengeance.

And what of the pioneers? Where are they now? Bob Moog is back to his first love in life, the theremin. He lives in Asheville, North Carolina, where he has a small company, Big Briar, that makes theremins. Bob and Shirleigh Moog separated after they moved to Ashville, and in 1996 Bob married Ileona Grams. After a long legal battle Bob recovered the right to use his own name, "Moog Music, Inc.," and now owns the registration of the trademark "Minimoog." He has started to make the Minimoog Voyager (a new version of the Minimoog) and also a line of effects modules called the Moogerfooger. After his adventures in the commercial world, Bob has returned to his roots, experimentation. He also loves to perform on the theremin, playing a remarkable duet with Keith Emerson on Moog synthesizer. He has also played with Don Buchla on stage in the Lincoln Center in a performance of Terry Riley's "In C."

Buchla too continues to design and manufacture synthesizers and space controllers ("Thunder" and "Lightning"). He still lives in San Francisco and still performs.

As individuals, Moog and Buchla don't seem so different. But how they interacted with the wider culture at crucial stages as they took their inventions forward made all the difference in the world. The reason we have a new instrument in the family of musical instruments is because Bob Moog was a boundary shifter, prepared to let the culture help shape his instrument. He listened to the users. In the end, he was more of a sixties guy than anyone ever realized.

Discography

Air, *Moon Safari* (Virgin, 1998)

Barron, Louis, and Bebe Barron, *Forbidden Planet* (Small Planet, 1989)

Bass, Sid, *Moog Espäna* (RCA, 1969)

Baxter, Les, *Moog Rock* (Crescendo, 1972)

Beach Boys, "Good Vibrations" (Capitol, 1966)

—— *Pet Sounds* (Capitol, 1966)

Beatles, *Sergeant Pepper* (Parlophone, 1967)

—— *Abbey Road* (Apple, 1969)

Beaver, Paul, and Bernie Krause, *The Nonesuch Guide to Electronic Music*
     (Elektra, 1968)

—— *Ragnarok: Electronic Funk* (Limelite, 1969)

—— *In a Wild Sanctuary* (Warner Brothers, 1970)

—— *Gandharva* (Warner Brothers, 1971)

Birtwistle, Harrison, *Triumph of Time: Chronometer* (Argo, 1975)

Blake, Tim, *Crystal Machine* (Egg, 1977)

Bley, Paul, *The Paul Bley Synthesizer Show* (Milestone, 1970)

Byrds, *The Notorious Byrd Brothers* (Columbia, 1968)

Can, *Tago Mago* (Mute, 1971)

Captain Beefheart and the Magic Band, *Safe as Milk* (Buddha, 1967)

Carlos, Walter, *Switched-On Bach* (Columbia, 1968)

—— *The Well-Tempered Synthesizer* (Columbia, 1968)

—— *Switched-On Bach II* (Columbia, 1973)

—— *Walter Carlos by Request* (Columbia, 1975); includes "Dialogues for Piano and Two Loudspeakers"

Carlos, Wendy, *Sonic Seasonings* (Columbia, 1972)

—— *Switched-On Brandenburgs* (Columbia, 1980)

Ciani, Suzanne, *Seven Waves* (Finnadar, 1982)

—— *Pianissimo II* (Seventh Wave, 1996)

Cream, *Disraeli Gears* (Atco, 1967)

Denny, Martin, *Exotic Moog* (Liberty, 1969)

Doors, "Light My Fire" (Elektra, 1967)

Doors, *Strange Days* (Elektra, 1967)

Droste, Keith, *Big Band Moog* (Realistic, 1970)

Emerson, Lake and Palmer, *Emerson, Lake and Palmer* (Atlantic, 1971); includes "Lucky Man"

—— *Tarkus* (Atlantic, 1971)

—— *Pictures at An Exhibition* (Atlantic, 1972)

Eno, Brian, *Music for Airports* (EG, 1978)

Fripp, Robert, and Brian Eno, *No Pussyfooting* (Island, 1973)

Garson, Mort, *The Zodiac Cosmic Sounds* (Elektra, 1967)

—— *The Wozard of IZ: An Electronic Odyssey* (A&M, 1969)

—— *Electronic Hair Pieces* (Mayfair, 1969)

—— *Black Mass/Lucifer* (MCA, 1971)

Gold, Marty, *Moog Plays the Beatles* (Avco, 1970)

Grateful Dead, *Anthem of the Sun* (Warner, 1968)

—— *Aoxomoxoa* (Warner, 1969)

Hambro, Leonid, and Gershon Kingsley, *Switched On Gershwin* (Avco, 1970)

Hammer, Jan, *Black Sheep* (Elektra, 1979)

Hankinson, Mike, *The Unusual Classical Synthesizer* (ABC, 1972)

Harrison, George, *Electronic Sounds* (Zapple, 1969)

Haskell, Jeff, *Switched-On Buck* (Capitol, 1971)

Hawkwind, *In Search of Space* (United Artists, 1971)

Henry, Pierre, *Experimental Music*, Vol. 2, *Le Voyage* (Limelight, 1968)

Hoskins, William, *Galatic Fantasy Eastern Reflections* (Spectrum, 1979)

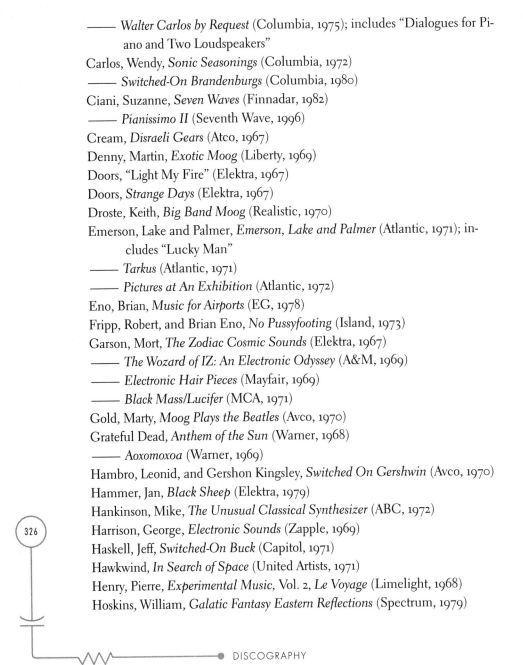

Hot Butter, *Popcorn* (Musicor, 1972)

Hyman, Dick, *The Age of Electronicus* (Command, 1972)

—— *The Synthesizer* (Command, 1973)

Jarre, Jean-Michel, *Oxygene* (Polydor, 1977)

Kazdin, Andrew, and Thomas Z. Shepard, *Everything You Always Wanted to Hear on the Moog* (Columbia, 1971)

Kingsley, Gershon, *First Moog Quartet* (Audiofidelity, 1969)

—— *Music to Moog By* (Audiofidelity, 1969)

Kraftwerk, *Autobahn* (Phillips, 1974)

—— *Trans-Europe Express* (Capitol, 1977)

—— *Computerworld* (EMI, 1981)

King Crimson, *Court of the Crimson King* (Island, 1969)

Light, Enoch, *Spaced Out* (Project 3, 1969)

Lothar and the Hand People, *Presenting—Lothar and the Hand People* (Capitol, 1969)

Mann, Sy, and Jean-Jacques Perrey, *Switched-On Santa: The Moog Synthesizer Plays the Merriest of Christmas Favorites* (Pickwick, 1970)

Melvoin, Mike, *The Plastic Cow goes MOOOOOOG* (Dot, 1969)

Montenegro, Hugo, *Moog Power* (RCA, 1969)

Moody Blues, *Days of Future Passed* (Decca, 1967)

Moog Cookbook, *The Moog Cookbook Plays the Classic Rock Hits* (Restless, 1997)

Moog Machine, *Switched-On Rock* (Columbia, 1969)

—— *Christmas Electric* (Columbia, 1970)

Mother Mallard's Portable Masterpiece Co, *Mother Mallard's Portable Masterpiece Co 1970–1973* (Cuneiform, 1999)

Nice, *Ars Longa Vita Brevis* (Immediate, 1969)

—— *Five Bridges Suite* (Mercury, 1970)

Oliveros, Pauline, *Alien Bog/Beautiful Soop* (Pogus, 1997)

Orb, *Adventures Beyond the Ultraworld* (Island, 1991)

Perrey, Jean-Jacques, *The Amazing New Electronic Pop Sound of Jean-Jacques Perrey* (Vanguard, 1967)

—— *Moog Indigo* (Vanguard, 1970)

Perrey, Jean-Jacques, and Gershon Kingsley, *The In Sound from Way Out* (Vanguard, 1966)

—— *Kaleidoscopic Vibrations: Spotlight on the Moog* (Vanguard, 1967)

Pink Floyd, *Piper at the Gates of Dawn* (EMI, 1967)

—— *Meddle* (EMI, 1971)

—— *Obscured by Clouds* (EMI, 1972)

—— *Dark Side of the Moon* (EMI, 1973)

Powell, Rick, *Switched-On-Country* (RCA, 1970)

—— *The Rick Powell Choir Book* (Word, 1974)

Powell, Roger, *Cosmic Furnace* (Atlantic, 1973)

Reich, Steve, *Early Works* (Nonesuch, 1987)

Riley, Terry, *In C* (Columbia, 1968)

Rolling Stones, *Their Satanic Majesty Requests* (Decca, 1967)

—— *Beggar's Banquet* (Decca, 1968)

Roxy Music, *Roxy Music* (Island, 1971)

—— *For Your Pleasure* (Island, 1973)

Rudin, Andrew, *Tragoedia* (Nonesuch, 1968)

Rundgren, Todd, *A Wizard, A True Star* (Bearsville, 1974)

Schulze, Klaus, *Irrlicht* (PDO Records, 1972)

Scott, Christopher, *Switched-On Bacharach* (Decca, 1969)

—— *More Switched-On Bacharach* (Decca, 1970)

Sear, Walter, *The Copper Plated Integrated Circuit* (Command, 1969)

Sear, Walter, and Richard Hayman, *Electronic Evolutions* (Command, 1969)

Simon, Paul, and Art Garfunkel, *Bookends* (Columbia, 1968)

Stereolab, *Switched-On Stereolab* (Too Pure, 1992)

Chris Stone, *Gatsby's World Turned On Joplin* (ABC, 1974)

Subotnick, Morton, *Silver Apples of the Moon* (Nonesuch, 1967)

—— *The Wild Bull* (Nonesuch, 1968)

—— *Touch* (Nonesuch, 1969)

Summer, Donna, *The Dance Collection* (Casablanca/Polygram, 1987); includes "I Feel Love"

Sun Ra, *My Brother the Wind* (Variety, 1969)

—— *My Brother the Wind*, Vol. 2 (Variety, 1970)

—— *The Solar-Myth Approach*, Vols. 1 and 2 (Variety, 1970)

Tangerine Dream, *Alpha Centauri* (Ear, 1971)

Tomita, *Snowflakes Are Dancing* (RCA, 1974)

Tonto's Expanding Headband, *Zero Time* (Embryo, 1971)

Trythall, Gil, *Nashville Gold Switched-On Moog* (Athena, 1969)

—— *Switched-On Nashville Country Moog* (Athena, 1970)

White, Ruth, *Short Circuits* (EMI, 1972)

White Noise, *An Electric Storm* (Island, 1971)

Who, *Who's Next* (Decca, 1971)

Who, "Won't Get Fooled Again" (Decca, 1971)

Williams, Mason, *The Mason Williams Ear Show* (Warner, 1968)

Wonder, Stevie, *Music of My Mind* (Motown, 1972)

—— *Talking Book* (Motown, 1972)

—— *Innervisions* (Motown, 1973)

—— *Fufillingness' First Finale* (Motown, 1974)

Wurman, Hans, *The Moog Strikes Bach* (RCA, 1969)

—— *Chopin à la Moog with Lots of Strings Attached* (RCA, 1970)

Zappa, Frank, *Uncle Meat* (Rykodisc, 1967)

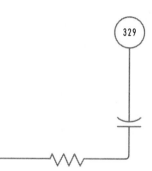

Sources

Unless otherwise indicated, quotations in the text were taken from interviews we conducted with the following people: Will Alexander (1-30-99), David Borden (5-3-96; 9-5-96), Don Buchla (5-20-96; 4-4-97), Leah Carpenter (6-22-98), Malcolm Cecil (3-31-00), John Chowning (4-4-97), Suzanne Ciani (8-23-98), David Cockerell (8-30-00), Herb Deutsch (4-19-97), Rachel Elkind (10-10-99), Keith Emerson (1-30-99), Linda Fisher (5-15-96), Ken Fung (4-24-96), Bill Hemsath (8-8-98), Danny Holland (6-22-98), Bernie Hutchins (4-18-96), Ikutaro Kakehashi (1-30-99), Brian Kehew (7-20-00), Gershon Kingsley (4-1-00), Bernie Krause (8-24-98), Bill Maginnis (7-11-00), Bob Margouleff (4-15-00), Bob Moog (6-5-96; 6-6-96; 11-15-97; 11-16-97), Tom Oberheim (4-4-97), Pauline Oliveros (12-2-00), Maggi Payne (12-2-00), Alan Pearlman (6-8-00), Jay Pollock (3-15-97), Don Preston (9-3-96; 4-5-97), Jim Scott (10-26-97), Walter Sear (6-19-98), Ramon Sender (7-11-00), David Van Koevering (1-30-99), Reynold Weidenaar (4-19-97), Jon Weiss (5-6-96), Robin Wood (8-28-00), and Peter Zinovieff (8-29-00). We talked with but did not formally interview Tom Constanten, Larry Fast, Dennis Houlihan, David Kean, Roger Luther, Steve Martin, Tim Orr, Tom Rhea, Mark Vail, Felix Visser, and David Hillel Wilson.

Throughout the book we also draw upon the following sources: interviews with Don Buchla, Bob Moog, and Vladimir Ussachevsky, housed at the Yale Oral History of American Music archive; comments made by Malcolm Cecil, Keith Emerson, Bob Margouleff, Bob Moog, and David Van Koevering at the "Keyboard Meets Modern Technology" panel and video conference, the Smithsonian Institution, April 14–16, 2000; interviews with Don Buchla, Ben Burtt, Wendy Carlos, Suzanne Ciani, Keith Emerson, Brian Eno, and Tom Oberheim quoted in Tom Darter and Greg Armbruster, eds., *The Art of Electronic Music* (New York: William Morrow, 1984; referred to in the notes as *AOEM*); the website Moogarchives.com.

In Chapter 5 we make extensive use of three books: Tom Wolfe, *Electric Kool-Aid Acid Test* (New York: Bantam, 1969); Carol Brightman, *Sweet Chaos: The Grateful Dead's American Adventure* (New York: Pocket Books, 1998), and Charles Perry, *The Haight-Ashbury: A History* (New York: Random House 1984). In Chapter 7 we draw on the website www.wendycarlos.com (all unreferenced quotes come from that source) and Arthur Bell's Playboy interview, "Wendy/Walter Carlos: A candid conversation with the 'switched-on Bach' composer who, for the first time, reveals her sex-change operation and her secret life as a woman," *Playboy*, May 1979.

A source for each chapter's epigraph is given in an unnumbered endnote at the beginning of the chapter's endnotes.

Notes

Introduction: Sculpting Sound

1. Paul Théberge, *Any Sound You Can Imagine: Making Music/Consuming Technology* (Hanover and London: Wesleyan University Press, 1997).
2. Bob Johnstone, "The Sound of One Chip Clapping," in *We Were Burning: Japanese Entrepreneurs and the Forging of the Electronic Age* (New York: Basic Books, 1998).
3. Thomas Rhea, "The Evolution of Electronic Musical Instruments in the United States" (Ph.D. diss., George Peabody College for Teachers, 1972).
4. Joel Chadabe, *Electric Sound: The Past and Promise of Electronic Music* (Saddle River, NJ: Prentice Hall, 1997), has a chapter on the development of synthesizers. Théberge, *Any Sound You Can Imagine*, also covers analog synthesizers in brief. For a detailed ethnography of French developments, see Georgina Born, *Rationalizing Culture: IRCAM, Boulez and the Institionalization of the Musical Avant-Garde* (Berkeley: University of California Press, 1995).
5. The term analog was applied only in hindsight—after digital came along.
6. One important later use was drum synthesizers—we do not cover these in this book.
7. Théberge, *Any Sound You Can Imagine*, pp. 52–53, notes that the American Music Conference first started tracking the sales of synthesizers as a separate category in 1973 in its annual statistics on the music instrument industry. About 7,000 synthesizers were sold in the United States that year at a value $8 million. This had risen to 24,000 per year by the end of the decade.
8. Mark Vail, ed., *Vintage Synthesizers*, 2nd ed. (San Francisco: Miller Freeman, 2000).
9. For an introduction to this field, see S. Jasanoff, G. Markle, J. Peterson, and T. Pinch, eds., *The Handbook of Science and Technology Studies* (Beverley Hills and London: Sage, 1994). For an influential account, see Bruno Latour, *Science in Action* (Cambridge: Harvard University Press, 1987). For a popularization of the S&TS approach see Harry Collins and Trevor Pinch, *The Golem: What You Should Know*

about Science, 2nd ed. (Cambridge: Canto, 1998), and Harry Collins and Trevor Pinch, *The Golem at Large: What You Should Know about Technology* (Cambridge: Canto, 2002).

10. For an influential analysis of how the rules of scientific method can be treated as sociological rules of practice, see H. M. Collins, *Changing Order: Replication and Induction in Scientific Practice,* 2nd ed. (Chicago: University of Chicago Press, 1992). The deep historical influence of society on science is argued in Steven Shapin, *A Social History of Truth: Civility and Science in Seventeenth-Century England* (Chicago: University of Chicago Press, 1994).

11. For classical music see Christopher Small, *Musicking: The Meanings of Performing and Listening* (Hanover and London: Wesleyan University Press, 1998); for dance music see Sarah Thornton, *Club Cultures: Music Media and Subcultural Capital* (Hanover and London: Wesleyan University Press, 1996); and for goa-trance dance music see Timothy D. Taylor, *Strange Sounds: Music, Technology and Culture* (New York: Routledge, 2002).

12. Ludwig Wittgenstein, *Philosophical Investigations* (Oxford: Blackwell, 1967).

13. We are particularly indebted to the new blend of sociology, gender, material practice, and musicology found in Théberge, *Any Sound You Can Imagine;* Steve Waksman, *Instruments of Desire: The Electric Guitar and the Shaping of Musical Experience* (Cambridge: Harvard University Press, 1999); Susan McClary, *Feminine Endings: Music, Gender and Sexuality* (Minneapolis: University of Minnesota Press, 1991); Susan McClary, "The Politics of Silence and Sound," afterword to Jacques Attali, *Noise: The Political Economy of Music* (Minneapolis: University of Minnesota Press, 1999), pp. 149–158; Simon Frith, *Performing Rights: On the Value of Popular Music* (Cambridge: Harvard University Press); Simon Frith and Andrew Goodwin, eds., *On Record: Rock, Pop, and the Written Word* (New York: Pantheon, 1990); and Antoine Hennion, "An Intermediary between Production and Consumption: The Producer of Popular Music," *Science, Technology & Human Values* 14 (1989). See also Howard Becker, *Art Worlds* (Berkeley: University of California Press, 1982), and Emily Thompson, *The Sounscape of Modernity* (Cambridge: MIT Press, 2002).

14. One of these silences comes from Wendy Carlos, who declined to be interviewed for this book.

15. Musicologists have developed a sophisticated language for describing music. But we puzzle over the gap between their descriptions and how sound and music are experienced by ordinary listeners and even musicians. Our approach has been influenced by sociologist Tia DeNora and her attempts to develop a phenomenology of how "ordinary people" listen to music; see Tia DeNora, *Music in Everyday Life* (Cam-

333

bridge, UK: Cambridge University Press, 2000); and by the phenomenologist Don Ihde, *Listening and Voice: A Phenomenology of Sound* (Athens: Ohio University Press, 1976). The need to attend to ordinary practice in music has been emphasized by David Sudnow, *The Ways of the Hand: The Organization of Improvised Conduct* (Cambridge: Harvard University Press, 1978).

1. *Subterranean Homesick Blues*

Epigraph from Bob Dylan, "Subterranean Homesick Blues," 1965.

1. Bob's dad had one of the earliest call signs, W2FP. For the history of amateur radio in America, see Susan J. Douglas *Inventing American Broadcasting, 1899–1922* (Baltimore: Johns Hopkins University Press, 1987), pp. 190–191. Douglas notes the gender-based nature of this hobby. For comparison with other male subcultures, including synthesizer users, see Théberge, *Any Sound You Can Imagine,* pp. 131–138.

2. Albert Glinsky, *Theremin: Ether Music and Espionage* (Urbana and Chicago: University of Illinois Press, 2000).

3. Robert Moog, "The Theremin," *Radio and Television News* 51 (January 1954): 37–40.

4. Glinsky, *Theremin,* p. 284.

5. Interview with Bob Moog, *Keyboard,* February 1995, p. 96.

6. The synthesizer was built by Harry Olsen and Herbert Belar in 1957 at RCA's Sarnoff Laboratories in Princeton. It had 750 vacuum tubes and was programmed from punched paper tape rolls.

7. Robert A. Moog, "A Transistorized Theremin," *Electronics World* 65 (January 1961): 29–32, 125.

8. It was an exciting time in New York for experimental music: not only was John Cage active, but one of the key figures in the emergence of minimalism, La Monte Young, formed The Dream Syndicate in 1963. The Welshman John Cale, later of the Velvet Underground, was a member of this group, which played dronelike music (Terry Riley was also a member). Young was part of the downtown art and music crowd called Fluxus spearheaded by Yoko Ono and George Maciunas.

9. See, for instance, L. Hiller, "An Integrated Electronic Music Console," *Journal of the Audio Engineering Society* 13 (1965): 142.

10. The voltage-controlled oscillators were built around a "relaxation" oscillator. The output shape of the waveform is a sawtooth. In addition to the oscillator circuit itself, the key elements of the voltage-controlled oscillators were an adder for summing the input voltages into the oscillator, an exponential converter for converting the linear summed output into an exponential output, and wave shapers for converting the sawtooth output into triangle, sine, or pulse waveforms.

11. Additive synthesis is the process of adding sine waves together to make musical tones.

12. Moog recalls reading about a modular system in an earlier article by Harald Bode. Deutsch recalls that Moog was particularly influenced by Bode's Melochord. Other voltage-controlled instruments to be developed around this time were James Beauchamp's harmonic tone generator (see J. W. Beauchamp, "Additive Synthesis of Harmonic Musical Tones," *Journal of the Audio Engineering Society* 14 [1966]: 332–342), and Paul Ketoff's Synket. Neither of these instruments were put into commercial production. See Chadabe, *Electric Sound*, pp. 144–146, for the Synket, and P. Ketoff, "The Synket," *Electronic Music Review* 4 (October 1967): 39–41.

13. Le Caine designed an early performance synthesizer, the electronic sackbut; see Gayle Young, *The Sackbut Blues: Hugh Le Caine, Pioneer in Electronic Music* (Ottawa: National Museum of Science and Technology, 1989). On the failed attempt to market the sackbut in 1970, see Théberge, *Any Sound You Can Imagine*, pp. 49–51.

14. Jacqueline Harvey later became Moog's PR person in New York City.

15. R. A. Moog, "Voltage-Controlled Electronic Music Modules," *Journal of the Audio Engineering Society* 13 (1965): 200–206.

16. Nikolais's Moog synthesizer is today at the Museum of Musical Instruments at the University of Michigan, Ann Arbor.

2. Buchla's Box

Epigraph from Don Buchla, interview with authors, May 21, 1996.

1. Suzanne Ciani, *Seven Waves* (1982), and *Pianissimo II* (1996).

2. Sender was a student of Darrius Milhaud who had used records to experiment with vocal and pitch transformations during the 1920s.

3. Steve Reich and Jon Hassell were in the ensemble for the first performance of "In C" (accompanied by a light show).

4. Mark Vail, *Keyboard*, October 1992, p. 46.

5. There had been many previous experiments with optical sources of sound, the best known being the Russian Yevgeny Sholpo's variaphone, invented in 1932.

6. Subotnick quoted in Vail, *Keyboard*, p. 46.

7. Quoted in Chadabe, *Electric Sound*, p. 147.

8. As a student Moog had visited Raymond Scott's studio on Long Island, where the jazz musician and inventor had set up something that resembled a synthesizer, even incorporating an electromechanical version of an early sequencer. Scott gave Moog, who was twenty at the time, a job building a subassembly for his clavivox instrument, and Moog continued building units for Scott throughout the early 1960s.

335

9. Gene Zumchak, a Moog engineer, developed the Moog sequencer in 1968, but the idea was based on what Buchla had done. Mark Prendergast, *The Ambient Century: From Mahler to Trance—The Evolution of Sound in the Electronic Age* (New York: Bloomsbury Publishing, 2000), p. 84, claims that Raymond Scott credited Bob Moog with the first use of the word "sequencer."

10. Pauline Oliveros does not herself recall the Yankee Doodle incident.

11. Riley's own form of keyboard music (he later used the Prophet 5 synthesizer) meant that he did not use the Buchla on any of his famous recordings.

12. Marl Vail, *Keyboard*, October 1992, p. 50.

13. A third type of voltage is trigger voltage, which triggers events such as an envelope generator.

14. On the Buchla 100 the noise source was known as the white noise generator.

15. Bernie Hutchins's interview with Bob Moog, *Electronotes* 6, no. 45 (1974): 5.

16. Subotnick went on to make two more acclaimed albums, *The Wild Bull* (1968) and *Touch* (1969).

17. Quoted in Chadabe, *Electric Sound*, p. 148.

18. Memos and letters at the time written by Moog's new sales manager, Al Padorr, reveal that Padorr was particularly impressed by the Buchla-CBS demonstrations.

19. Buchla quoted in Chadabe, *Electric Sound*, p. 148.

20. The best known Buchla synthesizers are the Buchla 100, 200, 400, the Touché, and the Music Easel. He has also developed two MIDI controllers, Thunder and Lightning, and an electronic version of the marimba. Many of the Buchla synthesizers are on display at the Audities Foundation, Chinook Keyboard Center, Calgary.

21. Bernie Hutchins's interview with Bob Moog, *Electronotes* 7, no. 50 (1975): 9.

3. Shaping the Synthesizer

Herb Deutsch, "The First Moog Synthesizer," excerpt from NAHO, fall 1981, New York Sate Museum, The State Education Department. Available at www.Moogarchives.com.

1. Reynold Weidenaar, *Magic Music from the Telharmonium* (Metchuen, NJ: Scarecrow Press, 1995).

2. Laurens Hammond, like Bob Moog, was a Cornell alumnus (he graduated in mechanical engineering in 1917). See Mark Vail, *The Hammond Organ* (San Francisco: Miller Freeman, 1997). Hammond worked closely with W. L. Lahey, a church organist, in the development of his instrument. See Théberge, *Any Sound You Can Imagine*, p. 45.

3. Goebbels followed new developments in music technology and was very keen on the sound of the trautonium. He advocated regular trautonium concerts on the radio

for propaganda purposes. See Hans-Joachim Braun, "Technik im Spiegel der Musik des fruehen 20. Jahrhunderts," *Technikgeschichte* 59 (1992): 109–131.

4. Karin Bijsterveld, "A Servile Imitation: Disputes about Machines in Music, 1910–1930," in Braun, '*I sing the Body Electric*," pp. 121–134.

5. *Time*, November 4, 1966, p. 44.

6. The Ondes Martenot and trautonium both used a form of ribbon controller to produce continuous changes in pitch.

7. Vail, ed., *Vintage Synthesizers*, p. 121.

8. Ibid.

9. See, Waksman, *Instruments of Desire*, p. 188.

10. Patent No. 3,475,623, "Electronic high-pass and low pass filters employing the base to emitter diode resistance of bipolar transistors."

11. The ladder of paired transistors with capacitors in between permits low-frequency signals to pass up the ladder unattenuated, since the reactance of the capacitors is highest at low frequencies. At higher frequencies the signal is shunted around the emitter-emitter input and is sharply attenuated.

12. The cut-off slope is 24 dB/octave, which means that frequencies one octave above the cut-off point are reduced 24 dB over those at the cut-off frequency.

13. Quoted in Steve Smith, "The Best Filter Built," *Imooginations* (Chicago: Norlin, 1976), p. 10. Our description of the operation of the Moog filter is based upon this article.

14. A good example is Malcolm Cecil's and Bob Margouleff's track "Cybernaut" on *Zero Time* (1971).

15. At the 1966 AES fall convention Moog presented in the exhibit hall his "System A" studio synthesizer, *Journal of the Audio Engineering Society* 14 (1966): 364.

16. The Hoover is also one of the few inventions whose name onomatopoeically suggests the sound of the invention.

17. Dusan Bjelic is an acquaintance of one of the authors. When he applied for U.S. citizenship, it was recommended that he change his name to Larry Smith!

18. The other Japanese synthesizer giant, Korg, has a similar made-up name. Korg's founder, Tsutomu Katoh, was a nightclub proprietor who teamed up with Tadahi Osanai, an engineer and noted Japanese accordionist, to form Keio Electronic Laboratories (representing the combination of the first initials of Katoh's and Osanai's names). In 1967 they produced their first organ with programmable voices and adopted the company name Korg, a combination of the words Keio and Organ.

19. Moog's 1965 "Ultra-Short Form Catalog of Electronic Composition Instruments" lists a VCO, 901; a VCA, 902; a white sound source, 903; a filter, 904; a reverberation

337

unit, 905; a fixed filter bank, 907; an envelope generator, 911; a keyboard controller, 950; and a linear controller, 955, as well as other subsidiary modules.

20. R. A. Moog Co., "Short Form Catalog," 1967. Moog also offered a complete range of 900-series modules and other devices, including a frequency shifter and ring modulator made under license from Harald Bode.

21. Ibid.

4. The Funky Factory in Trumansburg

Epigraph from David Borden, interview with authors, May 3, 1996.

1. Jon F. Szwed, *Space in the Place: The Lives and Times of Sun Ra* (New York: Da Capo Press, 1998).

2. Quoted in Waksman, *Instruments of Desire*, p. 246.

3. Bill of Sales, R. A. Moog to Blair Printing Co., June 1, 1965.

4. Letter, Walter Sear to Bob Moog, January 19, 1966.

5. Ibid., February 7, 1966.

6. Letter, Bob Moog to Walter Sear, February 17, 1966.

7. Ibid., November 10, 1965.

8. Letter, Walter Sear to Bob Moog, July 14, 1966.

9. Letter, Bob Moog to Walter Sear, November 8, 1966.

10. David Revill, *The Roaring Silence John Cage: A Life* (New York: Arcade, 1992).

11. Ibid., p. 213. See also Michael Nyman, *Experimental Music: Cage and Beyond*, 2nd ed. (Cambridge: Cambridge University Press, 1998).

12. Herbert A. Deutsch, "A Seminar in Electronic Music Composition," *Journal of the Audio Engineering Society* 14 (1966): 30–31.

13. For an excellent review of such magazines and their history and place in the later industry, see Théberge, *Any Sound You Can Imagine*, pp. 93–130. Another publishing venture significant for the future of the synthesizer industry was *Electronotes*, started in 1968 by Bernie Hutchins, a technician in the Cornell University Electrical Engineering Department. This newsletter contains circuits, other technical information, and also interviews (including one with Moog). *Electronotes* was read by all the engineers we interviewed.

14. Letter, Bob Moog to Walter Sear, February 10, 1967.

15. Ibid., November 25, 1966.

16. Ibid., February 10, 1967.

5. Haight-Ashbury's Psychedelic Sound

Epigraph from Tom Wolfe, *Electric Kool-Aid Acid Test* (New York: Bantam, 1969), p.351.

1. Charles Perry, *The Haight-Ashbury: A History* (New York: Random House 1984), p. 36.
2. Ibid., p. 9.
3. On the link between psychedelic music and light shows, see Sheila Whiteley, *The Space between the Notes: Rock and the Counter-Culture* (London and New York: Routledge, 1992). Stereo FM radio was also important; see Susan Douglas, *Listening In: Radio and the American Imagination, from Amos 'n' Andy and Edward R. Murrow to Wolfman Jack and Howard Stern* (New York: Random House, 1999).
4. Perry, *The Haight-Ashbury*, p. 13.
5. Ibid., p. 13.
6. Carol Brightman, *Sweet Chaos: The Grateful Dead's American Adventure* (New York: Pocket Books, 1998), p. 43.
7. Perry, *The Haight-Ashbury*, p. 34.
8. Wolfe, *Electric Kool-Aid Acid Test*, pp. 124, 129.
9. Perry, *The Haight-Ashbury*, p. 45.
10. Ibid., p. 47.
11. Gene Anthony was the photographer; see, Brightman, *Sweet Chaos*, p. 53.
12. Wolfe, *Electric Kool-Aid Acid Test*, p. 223.
13. Brightman, *Sweet Chaos*, p. 53.
14. Wolfe, *Electric Kool-Aid Acid Test*, pp. 351–352.
15. Perry, *The Haight-Ashbury*, p. 94. The fear of nuclear destruction was a theme echoed in San Francisco bands at the time such as in the Grateful Dead's "Morning Dew" (composed by Tim Rose) and in the Quicksilver Messenger Service's "Pride of Man" (composed by Hamilton Camp). The German group Can opened the track "Oh Yeah" on their highly influential *Tago Mago* (1971) with a simulated nuclear explosion.
16. Brightman, *Sweet Chaos*, p. 101.
17. Ibid., p. 100. David Kean of the Audities Foundation holds these modules, and we are grateful to him for drawing them to our attention.
18. Tom Constanten, *Between a Rock and Hard Places: A Musical Autobiodyssey* (Eugene: Hulogosi, 1998), p. 27.
19. Tom Constanten remembers that he treated Jerry Garcia's voice through a Moog synthesizer. The track "Rosemary" features a heavily distorted voice with phasing and filtering, and "What's Become of the Baby" (which takes up most of Side Two) has vocals treated, distorted, and phased.
20. An important part of these developments was the evolution in sound recording technology; see Andre Millard, *America on Record: A History of Recorded Sound* (Cam-

339

bridge, UK: Cambridge University Press, 1995), and David Morton, *Off the Record: The Technology and Culture of Sound Recording in America* (New Brunswick: Rutgers University Press, 2000).

21. Part of the early Pink Floyd sound was achieved by the Binson Echorec effects box until they first used a VCS3 synthesizer in 1971. The year 1967 also saw the release of the Beatles' "Strawberry Fields Forever," Donovan's "Mellow Yellow" and "Sunshine Superman," Jimi Hendrix's "Purple Haze," Procol Harum's "A Whiter Shade of Pale," and Pink Floyd's "See Emily Play."

22. For instance, British psychedelia took on a much more pastoral and whimsical feel; see Simon Reynolds, "Back to Eden: Innocence, Indolence and Pastoralism in Psychedelic Music, 1966–1996," in Antonio Melechi, ed., *Psychedelia Britannica* (London: Turnaround, 1997), pp. 143–165.

23. Steve Jones, *Rock Formation: Music Technology and Mass Communication* (Newbury Park: Sage, 1992), is one of the few books to draw attention to the importance of technology in a cultural studies analysis.

6. An Odd Couple in the Summer of Love

Epigraph from Bob Dylan, "ballad of a Thin Man," 1965.

1. Bernie Krause, *Into a Wild Sanctuary* (San Francisco: Heyday Press, 1998).

2. Jac Holzman and Gavin Daws, *Follow the Music: The Life and High Times of Elektra Records in the Great Years of American Pop Culture* (Santa Monica: First Media, 1998).

3. Her life formed the basis for the Burt Lancaster movie *Elmer Gantry*.

4. Quoted in Mark Cunningham, *Good Vibrations: A History of Record Production* (London: Sanctuary, 1998), p. 123.

5. Holzman and Daws, *Follow the Music*, pp. 98–99.

6. Nonesuch commissioned Morton Subotnick's acclaimed *Silver Apples of the Moon* 1967.

7. McGuinn is reputed to have been playing an arrangement of Bach's "Jesu, Joy of Man's Desire" on the banjo and got the intro to "Mr. Tambourine Man" from that. It sounds much more like the Beatles "What You're doing" from *Beatles for Sale* (1964).

8. One of the bonus tracks on the reissued CD (1997) is "Moog Raga," an instrumental credited to Roger McGuinn, who later bought his own Moog. McGuinn had intended to release a whole album of Moog material, but the idea proved not commercial enough to reach fruition.

9. Prendergast, *Ambient Century*, p. 230.

10. Holzman and Daws, *Follow the Music*, p. 204.

11. Ray Manzarek, *Light My Fire: My Life with the Doors* (New York: Putnam, 1998), pp. 256–257.

12. According to Bernie Krause (personal communication, March 13, 2001,) they couldn't get back to the original sounds "because everyone from Rothchild to the members of the group were so stoned at the session that there was nothing they could get back to."

13. Synthesists Bernie Krause, Don Preston, Patrick Gleeson, and Nyle Steiner were all involved on the final score. Preston used a perspex modular Moog and told us he was paid $2,000. For an account of the use of the synthesizer in *Apocalypse Now*, see Bob Moog, "Soundtrack to Apocalypse Now," in *AOEM*, pp. 260–267.

14. Harrison famously was involved in litigation as to the originality of his song "My Sweet Lord."

15. Mason Williams was a singer/songwriter famous for having written (with Nancy Ames) the hit single "Cinderella-Rockefella." There is one synthesizer track, "Generatah-Oscillatah" featuring the Moog—see *The Mason Williams Ear Show* (1968).

16. Earlier George Martin had bought a Moog synthesizer from Beaver and Krause. He provided liner notes for the Beaver and Krause album *Ragnarok: Electronic Funk* (1969).

17. The synthesizer on *Abbey Road* was probably set up by Mike Vickers of Manfred Mann—he was there in the studio at Abbey Road when the Moog was brought in (August 5, 1968), and he later showed Keith Emerson how to use his Moog (see Chapter 10).

18. For details see, Walter Everett, *The Beatles as Musicians: Revolver through the Anthology* (New York and Oxford: Oxford University Press, 1999), pp. 252–259.

19. Interview quoted from "One More for the Road," *Mojo*, October 2000, p. 71.

20. An *Electronic Studio Manual* for the Moog was produced in 1968-69 by Ronald Pellegrino, a music professor at Ohio State University.

21. Théberge, in *Any Sound You Can Imagine*, pp. 208–209, notes that these terms often link the sound to bodily sensations and there is often a curious reversal of social expectations concerning the value relationships attributed to any given pair, that is, a "fat" sound is preferred to a "thin" one.

22. As Bob Margouleff told us, "Stars twinkle, sounds trinkle."

23. *In a Wild Sanctuary* was pivotal for Krause because its "environmental impressionism" led to his later career in the world of bioacoustics.

24. *Gandharva*, a blend of electronics, jazz, rock, and gospel, includes jazz legends Gerry Mulligan, Howard Roberts, and Bud Shank and featured other artists like Leroy Vinnegar, Mike Bloomfield, and Gale Laughton.

341

7. *Switched-On Bach*

Epigraph from Glenn Gould, "More Notes by Glenn Gould and Wendy Carlos on the Well-Tempered Synthesizer," *Switched-On Boxed Set*, "Book Two: Original Notes," 1999, p. 20. Gould himself earlier had a controversial but hugely successful classical hit with Goldberg Variations.

1. This Columbia Masterworks album was based on the book *Rock and Other Four Letter Words* by J. Marks.
2. *AOEM*, p. 124.
3. Ibid., p. 299.
4. Ibid., pp. 122–123.
5. Ibid., p. 122.
6. Phillip Ramey, *Sonic Seasonings* (1972), "Walter Carlos: Then, Now and In-Between." Trans-Electronic Music Productions, Inc.
7. Wendy Carlos, "Evolution of a Recording Studio," *Switched-On Boxed Set*, "Book One: New Notes," 1999, p. 41.
8. *AOEM*, pp. 121–122.
9. Arthur Bell, "Wendy/Walter Carlos: A candid conversation with the 'switched-on Bach' composer who, for the first time, reveals her sex-change operation and her secret life as a woman," *Playboy*, May 1979, p. 90.
10. Judith Butler, *Bodies That Matter: On the Discursive Limits of Sex* (New York: Routledge, 1993).
11. *AOEM*, p. 122.
12. "Dialogues for Piano and Two Loudspeakers" (1963); "Episodes for Piano and Electric Sound" (1964); "Variations for Flute and Electronic Sounds" (1964); "Episodes for Piano and Tape" (1964); "Pompsities for Narrator and Tape" (1965); and "Noah" (1965).
13. *Playboy*, May 1979, p. 83.
14. There is a long history of virtuoso performers offering their own interpretations of masterworks—conductors and pianists make their careers from this very aspect of the business. Leopold Stokowski had already brought out an orchestrated version of Bach's "Toccata and Fugue in D Minor," which was set to animation in Disney's *Fantasia*.
15. *Playboy*, May 1979, p. 75.
16. Tod Dockstader, "Silver Apples of the Moon," *Electronic Music Review*, July 7, 1968, p. 32.
17. Walter Carlos, "Silver Apples of the Moon," *Electronic Music Review*, July 7, 1968, pp. 38–39.

18. Bob Moog, "Bob Moog Comments," *Switched-On Boxed Set,* "New Notes," 1999, p. 8.
19. *AOEM,* p. 130.
20. Susan Reed, "After a Sex Change and Several Eclipses, Wendy Carlos Treads a New Digital Moonscape," *People,* July 1, 1985, p. 83.
21. *Playboy,* May 1979, p. 100.
22. *AOEM,* p. 124.
23. *Journal of the Audio Engineering Society* 18, no. 1 (February 1970): 69.
24. Susan Reed, "After a Sex Change," p. 83.
25. Wendy Carlos continues to influence generations of synthesizer groups, from the Human League through to Stereolab.
26. Interview with Bob Moog, *Plug,* Fall 1974, p. 2.
27. Krause, *Into a Wild Sanctuary,* p. 54.
28. See James P. Kraft, *From Stage to Studio: Musicians and the Sound Revolution, 1890–1950* (Baltimore: Johns Hopkins University Press, 1996).

8. *In Love with a Machine*

Epigraph from Bob Moog, interview with authors, June 5, 1996.
1. *AOEM,* p. 252.
2. Chadabe, *Electric Sound,* p. 86.
3. Two examples of compositions produced with this method are "Bye, Bye Butterfly" (1965) and "1 of 4" (1966).
4. *AOEM,* pp. 252–253.
5. Ibid., p. 253.
6. Prendergast, *The Ambient Century,* p. 135.
7. *AOEM,* p. 257.
8. This account supports Susan McClary's suggestion that different genres of music are differently gendered; see *Feminine Endings.*
9. Evelyn Fox Keller, *A Feeling for the Organism: The Life and Work of Barbara McClintock* (New York: W. H. Freeman, 1983).
10. This new sort of hybrid identity is akin to the cyborg identities described by Donna Haraway, *Symians, Cyborgs and Women: The Reinvention of Nature* (New York: Routledge and Kegan Paul, 1991).

343

9. *Music of My Mind*

Epigraph quoted in Martin E. Horn, *Innervisions: The Music of Stevie Wonder* (Bloomington: 1stBooks Library, 2000), p. 110.

1. The synthesists they planned to include in their Expanding Head Band were Paul Beaver and Bernie Krause, Walter Carlos, Walter Sear, and Ruth White, another pioneering woman synthesist (see her 1972 recording *Short Circuits*).
2. He was also influenced by Marvin Gaye's *What's Going On* (1971).
3. These albums were produced in a number of studios, including: Electric Lady and Media Sound in NYC; Crystal Industries, Westlake Audio, and the Record Plant in LA; and Air Studios in London; see Horn, *Innervisions*, pp. 109–161.
4. The Unofficial Stevie Wonder Internet Archive.
5. The tour helped make number 1 hits of two singles released within the next year: "Superstition" and "You are the Sunshine of My Life."
6. Other notable black performers who made a name for themselves as synthesists were Chick Corea and Herbie Hancock, who both played keyboards with Miles Davis before going on to solo success.
7. Carol Cooper, "The Soul Nation Climbs Aboard," in Ashley Kahn, Holly George-Warren, and Shawn Dahl, *Rolling Stone: The Seventies* (Boston: Little, Brown, 1998), pp. 44–47.
8. Horn, *Innervisions*, p. 157.
9. *Little Stevie Wonder: The 12 Year Old Genius* is the title of a 1963 album. Berry Gordy, Jr., renamed Steveland Morris Little as Stevie Wonder.
10. Patricia Romanowski and Holly George-Warren, *The New Rolling Stone Encyclopedia of Rock & Roll* (New York: Rolling Stone Press), p. 1091.
11. *Music of My Mind* (1973), CD program notes.
12. Mark Mothersbaugh (Devo), *Tonto Rides Again* (1996) CD program notes.

10. Live!

Epigraph from Keith Emerson, interview with authors, January 30, 1999.
1. Donal Henahan, "Is Everybody Going to the Moog?" *Sunday New York Times*, August 24, 1969.
2. Allen Hughes, "Moog Approves of Moog-Made Jazz," *New York Times*, August 29, 1969.
3. See David E. Nye, *American Technological Sublime* (Cambridge: MIT Press, 1999).
4. Even earlier Richard Teitelbaum had brought the first Moog synthesizer to Europe performing over 200 concerts with it and helping to found the pioneering live electronic music group Musica Electronica Viva in Rome in 1966. They were also one of the first users of the VCS3 synthesizer live. The Synket was the first synthesizer ever used in live performance in April 1965. John Eaton performed "Songs for RPB" for

soprano, Synket, and piano, accompanied by soprano Michiko Hirayama in a concert at the American Academy in Rome. Another important early user of the modular Moog live was the jazz musician Paul Bley, who took one on a tour of Europe in 1970 and performed one of the early live synthesizer concerts at Carnegie Hall.

5. Friedkin on his next movie, *The Sorcerer* (1977), used Tangerine Dream for the music.

6. A monophonic vacuum tube instrument.

7. "Space Age Musicmaker: Gershon Kingsley" (2000), http://www.spaceagepop.com/kingsley.html.

8. More details of Keith's early life and his bands can be found in George Forrester, Martyn Hanson, and Frank Askew, *Emerson, Lake and Palmer: The Show That Never Ends* (London: Helter Skelter, 2001).

9. *AOEM*, p. 139.

10. See Waksman, *Instruments of Desire*, pp. 252–257. For an earlier influential analysis of "cock rock," see Simon Frith and Angela McRobbie, "Rock and Sexuality," in *On Record*.

11. The Nice, *Ars Longa Vita Brevis* (1969).

12. Letter, Walter Sear to Tony Stratton-Smith, January 16, 1970, quoted in Vail, *Vintage Synthesizers*, p. 117.

13. The synthesizer programming for early ELP albums may also have been carried out by producer/engineer Eddy Offord. Offord recalls, "I was actually the one who programmed all the sounds for him"; see Richard Buskin, *Inside Tracks: A First-Hand History of Popular Music from the World's Greatest Record Producers and Engineers* (New York: Avon, 1999), p. 174.

14. According to Keith's keyboard technician Will Alexander, the "Lucky Man" solo was actually the square waves (not sine waves), very slightly out of tune, and a low-pass filter with a little resonance on it to get it to ring. Keith finds it hard to recall exactly how he did the solo because he is embarrassed about it, maintaining that "Lucky Man" was not a true representation of what ELP was all about. He told us once that before one tour he had had to write to *Keyboard* magazine for a copy of the solo because he had forgotten how to play it!

11. Hard-Wired—the Minimoog

Epigraph by Bill Hemseth, interview with authors, August 8, 1998.

1. The actual figures for modular systems shipped from Trumansburg show a big decline for 1970. Totals for different years were as follows (figures from Roger Luther

who has the complete shipping records): 1967: 23; 1968: 49; 1969: 99; 1970: 54; and 1971: 16.

2. Gershon Kingsley recalls Sun Ra coming to use his modular Moog studio system — Kingsley provided the programming. Sun Ra's use of the Minimoog can be found on *My Brother the Wind* (1969), *My Brother the Wind*, Vol. 2 (1970), and *The Solar Myth Approach*, Vols. 1 and 2 (1970). See Szwed, *Space Is the Place: the Lives and Times of Sun Ra.*

3. Interview with Jan Hammer by Tom Rhea, *Imooginations* II, (Lincolnwood, IL: Norlin Music, 1977), p. 9.

4. Bob Moog, "On Synthesizers: The Minimoog Era," *Keyboard*, August 1981, pp. 50–51.

5. Hammer in interview with Rhea, *Imooginations* II, p. 9.

6. Roger Powell in interview with Rhea, *Imooginations* II, p. 12.

7. Bob Moog, personal communication, August 21, 2001.

8. The solo is "Waka/Jawaka" on Frank Zappa and the Mothers of Invention, *Waka/Jawaka* (Bizarre, 1972).

9. Bob Moog, personal communication, August 21, 2001.

12. Inventing the Market

Epigraph from Bob Moog, interview with authors, June 5, 1996.

1. Lenny Dee was a famous Hammond organ player.

2. See Colin Clark and Trevor Pinch, *The Hard Sell: The Language and Lessons of Street-wise Marketing* (London: Harper Collins, 1995), for an analysis of how such techniques work on streets and markets.

3. In the nineteenth century these traveling salesmen were known as "drummers"; see Olivier Zunz, *Making America Corporate, 1870–1920* (Chicago and London: University of Chicago Press, 1990).

4. Moog kept 15 percent of the company, which was worth $300,000 by the time Norlin bought Moog Music in 1973.

5. The only other synthesizer associated with the Musonics name was an upgrade of the Sonic V known as the Sonic VI. This synthesizer was considered to be so unreliable it acquired the nickname the "chronic sick."

6. See Clark and Pinch, *The Hard Sell.*

7. Van Koevering's strategy seems to fit with Latour's analysis of Pasteur, moving back between field and lab as he enlisted new groups to the cause of his anthrax virus. Bruno Latour, "Give me a Laboratory and I will Raise the World," in Karin D.

Knorr-Cetina and Michael Mulkay, eds., *Science Observed: Perspectives on the Social Study of Science* (London and Beverly Hills: Sage, 1983), pp. 141–170.

13. Close Encounters with the ARP

Epigraph from Roger Powell (talking about a well-known 2600 user) in Vail, *Vintage Synthizers*, p. 128

1. Some of the more significant American companies were Oberheim, Linn, Sequential, Emu, Ensoniq, EML, and Serge.
2. Geoffrey Giuliano, *Behind Blue Eyes: The Life of Pete Townshend* (New York: Plume, 1997), p. 91.
3. *AOEM*, p. 223.
4. Bob Moog, "ARP 2600 Most Popular Modular Synth," in Vail, *Vintage Synthesizers*, pp. 124–125.
5. Interview with Bob Moog, *Keyboard*, February 20, 1995, p. 94.
6. *AOEM*, p. 96.
7. Craig Waters with Jim Aikin, "The Rise and Fall of ARP Instruments: Too Many Chefs in the Kitchen," in Vail, *Vintage Synthesizers*, pp. 48–58.
8. CBS did take over one last successful ARP product, the Chroma synthesizer, which was the first ARP to use microprocessors and is still a favorite among ARP enthusiasts for its huge range of sounds.
9. *AOEM*, p. 241.
10. Ibid., p. 257.

14. From Daleks to the Dark Side of the Moon

Epigraph from Leonard Cohen, "Last Year's Man," 1971.

1. The Dalek voice was created by Brian Hodgson; see Desmond Briscoe and Roy Curtis-Bramwell, *The BBC Radiophonic Workshop: The First 25 Years* (London: BBC, 1983), p. 107.
2. The composer was Ron Grainer.
3. Delia Derbyshire and Brian Hodgson later teamed up with David Vorhaus to form White Noise. Their first record, *An Electric Storm* (1971), became a cult classic.
4. Tristram Cary quoted in Chadabe, *Electric Sound*, p. 151.
5. Bob Moog had earlier made a similar customized touch-sensitive keyboard for Wendy Carlos, but it was not a standard item on the Moog synthesizer at the time.
6. Walter Carlos, "On Synthesizers," *The Last Whole Earth Catalog* (Harmonsworth, UK: Penguin Books, 1971), p. 331.

7. EMS also developed a pitch-to-voltage converter, a random-voltage generator, and the octave filter bank. Later they developed a very successful vocoder (designed by Tim Orr).

8. Cunningham, *Good Vibrations*, p. 205.

9. Quoted in Bud Doerscuk, ed., *Rock Keyboard* (New York: Quill/Keyboard, 1986), p. 55.

10. *AOEM*, pp. 222–223.

11. Hawkwind, *In Search of Space* (1971).

12. Nick Kent, "Hawkwind Fly as a Kite," *Frendz* 31 (July 14, 1972): 21.

13. For instance, Klaus Schulze, *Irrlicht* (1972); Tangerine Dream, *Alpha Centauri* (1971); and Kraftwerk, *Autobahn* (1974). The EMS vocoder was used by Kraftwerk on *Computer World* (1981).

14. Jean-Michel Jarre's *Oxygene* (1977) is often described as a significant influence on the techno trance music of the late 1990s.

15. Tristram Cary, "Electronic Music, Background to a Developing Art," *Audio Annual*, 1971, pp. 42–49.

16. Robin also produces the Soundbeam, a remarkably simple interface to help disabled people make music.

Conclusion: Performance

1. Bill Wyman, *Stone Alone: The Story of a Rock 'n' Roll Band* (New York: Viking, 1990).

2. Although TONTO (but not its sounds) makes a cameo appearance in Brian De Palma's *Phantom of the Paradise* (1974).

3. According to film director Horace Ove, *Performance* is the closest movie there is to describing the psychedelic experience; see Jonathon Green, *Days in the Life: Voices from the English Underground, 1961–71* (London: Heinmann, 1988), pp. 90–91.

4. Jagger's friend, the film producer Kenneth Anger, used Jagger's compositions on the Moog for his Californian homoerotic art-movie short *Invocation of My Demon Brother* (1968). On listening to the music for that movie, Jon recognized it as coming from the little patch he had set up for Mick. As he told us, "That's my sound."

5. This book is not able to do justice to the enormous influence of the German tradition. Mark Prendergast in *The Ambient Century* has good coverage of the main synthesizer artists, and Mark Vail in *Vintage Synthesizers* deals with the main German synthesizer manufacturers. See also the interviews with figures like Klaus Schulze, Edgar Froese, and Christoph Franke in *AOEM*.

6. The United States Court of International Trade, "CASIO Inc versus The United States," October 7, 1994. 1994 WL 548786.

7. Sheila Jasanoff, *The Fifth Branch: Science Advisers as Policymakers* (Cambridge: Harvard University Press, 1990), shows the importance of studying legal and quasi-legal contexts where matters of science and technology policy are settled.

8. Trevor Pinch and Karin Bijsterveld, "Breaches and Boundaries in the Reception of New Technology in Music" (unpublished).

9. Craig H. Roell, *The Piano in America: 1890–1940* (Chapel Hill: University of North Carolina Press, 1989) p. 40.

10. Roell, *The Piano in America*, p. 54.

11. Pinch and Bijsterveld, "Breaches and Boundaries."

12. Roell, *The Piano in America*, p. 39.

13. Victor W. Turner, *The Ritual Process: Structure and Anti-Structure* (Harmondsworth: Penguin 1969), p. 81.

14. It is this transformative property which distinguishes a liminal entity from a "boundary object"—see Susan Leigh Star and James R. Griesemer, "Institutional Ecology, 'Translations,' and Boundary Objects: Amateurs and Professionals in Berkley's Museum of Vertebrate Zoology, 1907–39," *Social Studies of Science* 19 (1989): 387–420. Our use of the term "liminal entity" tries to capture not only the property of crossing boundaries but also the ability to produce transformations in the process.

15. Tracy Kidder, *The Soul of a New Machine* (New York: Avon Books, 1982).

16. On the importance of standardization in science see Simon Schaffer, "Late Victorian Metrology and its Instrumentation: A Manufacture of Ohms," in Robert Bud and Susan Cozzens, eds., *Invisible Connections: Instruments, Institutions and Science* (Bellingham, WA: SPIE Optical Engineering Press, 1992), pp. 23–56. For standardization in technology, see Ken Alder, *Engineering the Revolution: Arms and Enlightenment in France, 1763–1815* (Princeton: Princeton University Press, 1997). On standardization of volts in electrical appliances, see Joseph O'Connell, "Metrology: The Creation of Universality by the Circulation of particulars," *Social Studies of Science* 23 (1993): 129–173.

17. See W. Bijker, *On Bikes, Bulbs and Bakelites* (Cambridge: MIT Press, 1995).

18. Robert Moog, "The Keyboard Explosion: Ten Amazing Years in Music Technology," *Keyboard* 11, no. 10 (October 1985): 36–48. Théberge, *Any Sound You Can Imagine*, chapter 4, shows how the "democratizing" of the synthesizer is underpinned by several parallel developments that mesh with an expanded market for keyboards in the 1980s.

19. For "interpretative flexibility" see Trevor Pinch and Wiebe Bijker, "The Social Con-

349

struction of Facts and Artifacts," *Social Studies of Science* 14 (1984): 339–441. See also the essays in Bijker, Hughes, and Pinch, *The Social Construction of Technological Systems*. For a classic study of how social assumptions get embedded in technologies, see Donald MacKenzie, *Inventing Accuracy: A Historical Sociology of Nuclear Missile Guidance* (Cambridge: MIT Press, 1990).

20. See M. Akrich, "The De-scription of Technological Objects," in Wiebe Bijker and John Law, eds., *Shaping Technology/Building Society* (Cambridge: MIT Press), pp. 205–224; Ronald Kline and Trevor Pinch, "Users as Agents of Technological Change: The Social Construction of the Automobile in the Rural United States," *Technology and Culture* 37 (1996): 763–795; David E. Nye, *Electrifying America: Social Meanings of New Technology 1880–1940* (Cambridge: MIT Press); Ronald Kline, *Consumers in the Country: Technology and Social Change in Rural America* (Baltimore and London: Johns Hopkins University Press, 2000); Hugh Mackay and Gareth Gillespie, "Extending the Social Shaping Approach: Ideology and Appropriation," *Social Studies of Science* 22 (1992): 685–716; Ruth Schwartz Cowan, "The Consumption Junction: A proposal for Research Strategies in the Sociology of Technology," in Bijker, Hughes, and Pinch, *The Social Construction of Technological Systems*, pp. 17–50; and Steve Woolgar, "Configuring the User: The Case of Usability Trials," in John Law, ed., *A Sociology of Monsters: Essays on Power, Technology and Domination* (London: Routledge, 1991), pp. 58–100.

21. The term "script" is due to Akrich, "The De-scription of Technological Artifacts." The term "configure" is due to Woolgar, "Configuring the User." For a fascinating treatment of users in the development of the personal computer, see Thiery Bardini, *Bootstrapping: Douglas Engelbart, Coevolution, and the Origins of Personal Computing* (Stanford: Stanford University Press, 2000).

22. This argument is made for digital technology by Théberge, *Any Sound You Can Imagine*.

23. For work on salespeople in the history of technology, see Olivier Zunz, *Making America Corporate, 1870–1920* (Chicago and London: University of Chicago Press, 1990); Claude S. Fischer, *America Calling: A Social History of the Telephone to 1940* (Berkeley: University of California Press, 1992); Carolyn Goldstein, "From Service to Sales: Home Economics in Light and Power, 1920–1940," *Technology and Culture* 38 (1992): 121–152; C. Cockburn and S. Ormrod, *Gender and Technology in the Making*; (Thousand Oaks and London; Sage 1973); and Kline, *Consumers in the Country*.

24. For a new view of markets from a Science and Technology Studies perspective, see Michel Callon, ed., *The Laws of the Markets* (Oxford: Blackwell, 1998).

25. This "mangling" of categories has been emphasized by Andrew Pickering, *The Mangle of Practice: Time, Agency, and Science* (Chicago: University of Chicago Press, 1995).

26. Pablo Boczkowski, in "Affording Flexibility: Transforming Information Practices in On-Line Newspapers" (Ph.D. diss., Cornell University, 2001), has suggested the term "boundary subject position" for such people and has drawn attention to their importance. Raghu Garud and Peter Karnøe similarly stress the importance of "boundary spanning" in entrepreneurship (indeed we can note the original meaning of *entre* as "between"). See Raghu Garud and Peter Karnøe, "Path Creation as a Process of Mindful Deviation," in Raghu Garud and Peter Karnøe, eds., *Path Dependence and Creation* (New Jersey and London: LEA, 2001), pp. 1–38. Another anthropological metaphor to capture the importance of these "in between zones and people" is the "trading zone"; see Peter Galison, *Image and Logic: A Material Culture of Microphysics* (Chicago: University of Chicago Press, 1997).

27. "Shift" denotes not only movement but also carries the semiotic meaning of shifting in and out of frames. It is also analogous to the "shape shifting" of the shaman.

28. Thus, sales people have turned out to be key actors in the synthesizer story—where would the synthesizer be without the activities of Walter Sear, Paul Beaver, Bernie Krause, David Van Koevering, and Tom Oberheim—all synthesizer salesmen at one time or another?

29. For work which draws attention to hybrids see, Bruno Latour, *We Have Never Been Modern* (Cambridge: Harvard University Press, 1993), and Haraway, *Symians, Cyborgs, and Women*.

30. The classic argument about the anti-technology element of the counterculture is to be found in Theodore Roszak, *The Making of a Counter Culture: Reflections on the Technocratic Society and Its Youthful Opposition* (New York: Doubleday, 1969). See also, Y. Ezrahi, E. Mendelsohn, and H. Segal eds., *Technology, Pessimism and Postmodernism* (Dordrecht: Kluwer, 1993).

31. Brightman, *Sweet Chaos*, p. 53, notes the influence of the counterculture upon the wider computer revolution. She claims that many of the key figures in Silicon Valley were present at counter-culture events like the Trips Festival.

32. Théberge, *Any Sound You Can Imagine*, p. 74. Casio sold 15 million instruments between 1980 and 1990.

33. Théberge, ibid., pp. 75–76, notes that this story has acquired mythical status and is more often told about Sequential's Prophet-5.

34. Théberge, ibid., points out that digital instruments lead to a much more abstract re-

lationship between the user and the machine. The knowledge of "patches" in the digital era is quasi-mathematical, abstract, and formal as opposed to the use of visual, aural, and tactile skills in the analog era.

35. There seems to be a degree of confusion as to what analog means. For some people analog means the interface—the knobs and wires. For others it means the actual technology producing the sound—analog technology being based on continuous variables rather than the discrete bits of the digital age. It is also worth noting that player-piano rolls provided a digital storage medium long before the analog days we write about here.

36. Greg Rule, *Electro Shock! Groundbreakers of Synth Music* (San Francisco: Miller Freeman, 1999), contains interviews with many modern synthesists who use analog equipment.

37. Digital emulations of old analog instruments are also now available. Such programs beg the question, however, because the sounds are produced digitally and the interface is invariably the keyboard and mouse—not the beloved knobs and patch wires of the analog era.

38. See, Fred Davis, *Yearning for Yesterday: A Sociology of Nostalgia* (New York and London: Free Press, 1979).

39. It is also worth bearing in mind that this analog revival encompasses pre-sixties technologies, like tube amplifiers, tube studios, the theremin, and the trautonium.

40. *AOEM*, p. 221.

41. For an analysis of the warmth of analog sound see Andrew Goodwin, "Sample and Hold: Pop Music in the Digital Age of Reproduction," *Critical Quarterly* 30 (1988): 34–49. Goodwin notes that in the early seventies, analog synths were held to be "cold" and "inhuman"; it was only later in the eighties that the analog sounds started to be called "warm."

42. Rap artists too sometimes prefer old equipment for its "dirty" distorted sound; see Tricia Rose, *Black Noise: Rap Music and Black Culture in Contemporary America* (Hanover, NH: University Press of New England, 1994). pp. 76–77.

43. For a massive work that traces the influences on "ambient music," see Mark Prendergast, *The Ambient Century* (with an exhaustive discography). Another approach can be found in the film *Modulations*. The follow-up book to the film contains several informative chapters and interesting interviews with today's electronic music artists: Peter Shapiro, ed., *Modulations: A History of Electronic Music: Throbbing Words on Sound* (New York: Caipirinha, 2000). Another exhaustive and intelligent examination is Simon Reynolds, *Energy Flash: A Journey through Rave Music and Dance Culture* (London: Macmillan, 1998). A set of amazing reflections

on electronic music can be found in DJ and writer Kodwo Eshwun's *More Brilliant than the Sun: Adventures in Sonic Fiction* (London: Quartet Books, 1999). The American Origins of techno are documneted in Dan Sicko, *Techno Rebels: The Renegades of Electronic Funk* (New York: Billboard Books, 1999).

44. Matthew Collin, *Altered State: The Story of Ecstasy Culture and Acid House*, 2nd ed. (London: Serpent's Tail, 1998), p. 67.

45. Ibid., p. 201.

46. Sheila Whiteley, "Altered Sounds," in Antonio Melechi, ed., *Psychedelia Britannica* (London: Turnaround, 1997), p. 139.

47. Ibid.

48. A more complete account would trace the influence of analog synth freak Rik Davis on the funk music of Juan Atkins, Kevin Saunderson, and Derrick May—the so-called Belleville Three (named after the small town outside Detroit where they all met). Atkins played with Davis (twelve years his elder and a former Vietnam vet) in Cybotron. Atkins went on to introduce Derrick May to electronic music through Yellow Magic Orchestra, Ultravox, Devo, the Human League, Gary Neuman, and Kraftwerk, which when combined with the funk of Parliament-Funkadelic and George Clinton led to techno. Derrick May famously described techno as "George Clinton and Kraftwerk caught in an elevator with only a sequencer to keep them company."

49. Collin, *Altered State*, p. 21.

Illustration Credits

Lyrics from Bob Dylan reprinted by permission. © Copyright 1965 by Warner Bros. Inc. © Copyright renewed 1993 by Special Rider Music. All Rights reserved. International copyright secured.

Lyrics from Leonard Cohen reprinted by permission of Stranger Management.

We acknowledge the following people and institutions for permission to use photographs:

Gene Anthony: 14, 15
Audio Engineering Society: 6 (*AES* 13, no. 1, June 1965), 8 (*AES* 18, no. 3, June 1970)
Malcolm Cecil: 25, 26, 27
Herb Deutsch: 5, 28
Elektra Entertainment Group: 19
R. A. Erdmann: 10
David Kean and Audities Foundation: 16 (permission to photograph Ken Kesey's Buchla Box), 35, 36, 37, 39, 40
Gershon Kingsley: 23, 31, 32
Bernie Krause: 17

Roger Luther of Moogarchives.com: 1, 9, 11, 18, 48
Robert Moog: 13, 43
MCA Records: 22
George Moog: 4
Jon Reis: 29, 30
Bill Reitzel: 7
David Van Koevering: 33, 34, 38, 41, 42, 44, 45, 46, 47
Jon Weiss: 2, 3
Lloyd M. Williams: 24 (from the 16 mm experimental movie *Rainbow's Children*)
Robin Wood and EMS: 53, 54, 55, 56

Glossary

CONTROLLER: A device (such as a keyboard, ribbon controller, or pitch wheel) that allows users to change a synthesizer's electrical circuits and thereby alter some aspect of the sound.

CONTROL VOLTAGE: An electrical signal that tells a voltage-controlled device (for example, a voltage-controlled filter or a voltage-controlled oscillator) to move to a different level or to change its parameters.

ENVELOPE: A control voltage that can be applied to any aspect of a sound (such as pitch, loudness, or brightness) to give it a distinctive shape or contour.

EXPONENTIAL CONVERTER: A device in which a parameter of the output increases exponentially relative to the input. For example, a two volt increase at the input might double the output frequency, or increase the amplitude of the output signal by a factor of four. In a linear device, by contrast, the output increases in direct proportion to the input.

FILTER: A device that allows particular frequencies of sound to pass through while removing other frequencies. A low-pass filter allows low frequencies to pass (reducing frequencies above its cutoff point), and a high-pass filter allows high frequencies to pass (reducing frequencies below its cutoff point). Moog's famous ladder filter is a low-pass filter named after the "ladder" of transistors in its circuit.

MIXER: A device used in live performance or studio recording which combines several channels of audio signals into one or more outputs. Also called a board.

MONOPHONIC SYNTHESIZER: A synthesizer capable of producing only one independently moving pitch at a time. In a monophonic synthesizer controlled by a keyboard, only one key can produce a pitch at a given time; a polyphonic synthesizer can produce more than one pitch at a given time.

OSCILLATOR: A device that produces sound through regularly repeating fluctuations in voltage.

PATCH: The combination of connections, settings, and adjustments that denotes a particular sound in a synthesizer.

PATCH CORDS: The cables used to connect synthesizer modules (inputs and outputs), similar in appearance to old telephone switchboard cables.

PITCH WHEEL: A small wheel to the left of the keyboard, sitting on its edge with about half the wheel protruding above the panel, used for pitch bending and vibrato.

PORTAMENTO: A smooth, uninterrupted glide in passing from one note to another.

POTENTIOMETER (POT): A device attached to a knob or slider that varies electrical signals on the synthesizer. In traditional radios and televisions, potentiometers are used for volume control.

RIBBON CONTROLLER: A device most often used for bending pitches; played by sliding the finger along a ribbon, which creates a varying electrical contact along a pair of thin longitudinal strips whose electrical potential changes from one end to the other.

RING MODULATOR: A special type of signal processor that accepts two signals as audio inputs and generates their sum and difference tones as the output (but not the original signals themselves), producing bell-like sounds because of the altered harmonic structure.

SAMPLE-AND-HOLD: A device that samples an incoming voltage to determine its level and then puts out a signal at that same level until the next time it is instructed to take a sample, while ignoring the intervening incoming voltages.

SEQUENCER: A device that continuously produces the same, predetermined arrangement of voltages. Depending on the part of the synthesizer to which it is applied, many different effects can be created (for example, a sequencer causes an oscillator to change pitch).

VOLTAGE-CONTROLLED AMPLIFIER (VCA): A device that alters the strength of the signal in proportion to a control voltage.

VOLTAGE-CONTROLLED FILTER (VCF): A filter whose cutoff frequency varies in proportion to a control voltage.

VOLTAGE-CONTROLLED OSCILLATOR (VCO): An oscillator whose frequency varies in proportion to a control voltage.

VOLT-PER-OCTAVE: A system developed by Bob Moog in which a one-volt change applied to a control input of a voltage-controlled oscillator produces a one-octave change in the pitch.

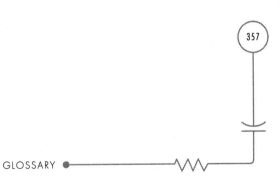

Index

INDEX

367

INDEX